SEARCHING FOR AFRICAN PROSPECTS

SEARCHING FOR AFRICAN PROSPECTS

Life as a Mining Engineer in Nigeria and Angola

Diana Chads

The Radcliffe Press

LONDON • NEW YORK

Published in 2006 by The Radcliffe Press
6 Salem Road, London W2 4BU

In the United States and in Canada
distributed by Palgrave Macmillan, a division of St Martin's Press
175 Fifth Avenue, New York NY 10010

ISBN 1 84511 182 6
EAN 978 1 84511 182 3

A full CIP record for this book is available from the British Library
A full CIP record for this book is available from the Library of Congress

Library of Congress Catalog card: available

Typeset in Sabon by Oxford Publishing Services, Oxford
Printed and bound in Great Britain by MPG Books Ltd, Bodmin

Contents

List of illustrations

Glossary

adobe	sun-dried clay brick used as a building material
alkali	local Nigerian judge
baraki	mining camp
barraque	men's housing quarters in camp
batook	special form of African drum
biltong	strips of dried raw meat
bushcow	short-horned West African buffalo
capataz	foreman
chef de poste	labour recruiter
chop	food
chupoya	portable bed
fuba	crushed mealies (local staple diet)
leat	trench or channel for conveying water
mallam	a man educated in Koranic studies
materiales de guerre	war materials
palaver	argument (in West African context) or discussion
sala	Muslim religious festival
seraki	king
shengie	hut
veld	open country
waca	clumsy
zareba	thorn gazebo

Foreword by G. C. Morrison

Will Chads was a great character, of that there can be no doubt. Essentially over many years our relationship was that of master and servant. As his stockbroker, he would walk into my office unannounced and unexpected, frequently when there was absolute bedlam with telephones ringing all around, and stand silently in the doorway until someone noticed him – a tall commanding figure in a rather crumpled beige lounge suit that must have seen many years service in the tropics, a smile on his face and a silver-topped cane completed the picture.

On one occasion when he looked particularly dishevelled, I asked him if next time he looked in the office, he would wear a better suit so that the other stockbrokers in the building would not think that we were rotten stockbrokers! He laughed, as I knew he would.

On being asked to sit down, there would follow a few minutes of banter. He was known as Captain Chads to the older members of our staff. On his personal effects I once noticed that a box had Major Chads printed on the side but such was his commanding presence, I never had the courage to ask why.

During the years I knew him, he had what seemed to me to be a well balanced life – winter indulging his love of opera and summers spent with his interest in tennis – rumour had it that in his younger years he played for his county. Certainly, I had many happy days at Wimbledon with him in the best seats and only once was payment requested – to pump up the tyres of his elderly but distinguished Daimler. He was a very generous man and I went with him on several occasions to the opera – again in the best seats. On the last occasion he failed to parade at the last moment and his daughters explained that he was unwell and sadly I never saw him again.

He will be remembered with the greatest possible affection by all who knew him.

G. C. Morrison

Introduction

William was known to all his friends as Captain Chads, only his close relatives knew him as Bill. He was a real English gentleman and wherever he was he always maintained his high standards of language, hygiene and behaviour. He was tall, handsome and always spoke in a friendly and kind manner.

William was born on 13 February 1893 when his father was a senior army officer stationed in Brisbane, Australia. The family returned to England and in September 1907, when he was 14 years' old, he entered Cheltondale College, Cheltenham. William excelled in sports and played hockey, football, rugby and cricket in his house teams, often as captain. He excelled in fencing, was a good tennis player and enjoyed boxing. He also became a prefect.

On leaving school he joined the army and started training at the Royal Military Academy at Woolwich. He was appointed second lieutenant, special reserve of officers at the Cork Royal Garrison Artillery on 10 October 1913, aged 20. From 11 August 1914 he joined the land forces in the royal horse and royal field artillery and soon found himself in France during the bitter fighting of the First World War. He particularly enjoyed serving as aide-de-camp to General Waldrop for whom he had great respect. His work often involved a high degree of bravery, especially when involved in reconnaissance, and this was rewarded on 5 May 1916 when he was awarded the Military Cross.

William was wounded during the battle and had to be treated by the medical team. His wounds were such that they had an enduring effect. Having completed a probationary staff course, he was sent to the main staff course with the 58th division, RFA on 15 October 1918.

In 1920 he was with the troops camped on the outskirts of Constantinople. Although it was already two years since the end of the war travel restrictions were still severe. William had to travel home on his own and special arrangements were made for the journey. On 4 September 1920 the British high commission in Constantinople (Istanbul) issued a document to William permitting him to travel over land to Belgrade to England for one journey only. It requests that 'all Allied Naval Military and other authorities to whom this pass may be presented will afford the bearer all reasonable facilities'. The French issued a pass to travel without stopping via Greece, Bulgaria, Romania, Serbia, Italy, Switzerland and France.

William now had freedom to explore the world and no matter how tough the going he carried out his intentions. His early sojourn to ski and skate in the cold mountains of Norway only served to whet his appetite for more exciting adventures to follow. He wrote frequent letters home giving vivid descriptions of the scenery and people he met and this discipline he continued throughout his excursions.

William was a keen photographer and his pictures of South America and Africa, together with his notes, give a vivid impression of the places he visited and all the dangers, thrills and spills he encountered.

After spending several years mining and surveying in Chile and Bolivia, mostly on foot and horseback, he had to start looking elsewhere for work and excitement. The next opportunity to work and explore a bit more of the world was when he was engaged by a mining company working in Nigeria followed by another company whose interests were in Angola where the wild life was particularly rich but food was extremely short. Survival was all important and often William could only keep himself and his men alive by accurate use of his gun. He only shot animals and birds for food and preferred the shot of the camera to the shot of the gun. Now the story begins.

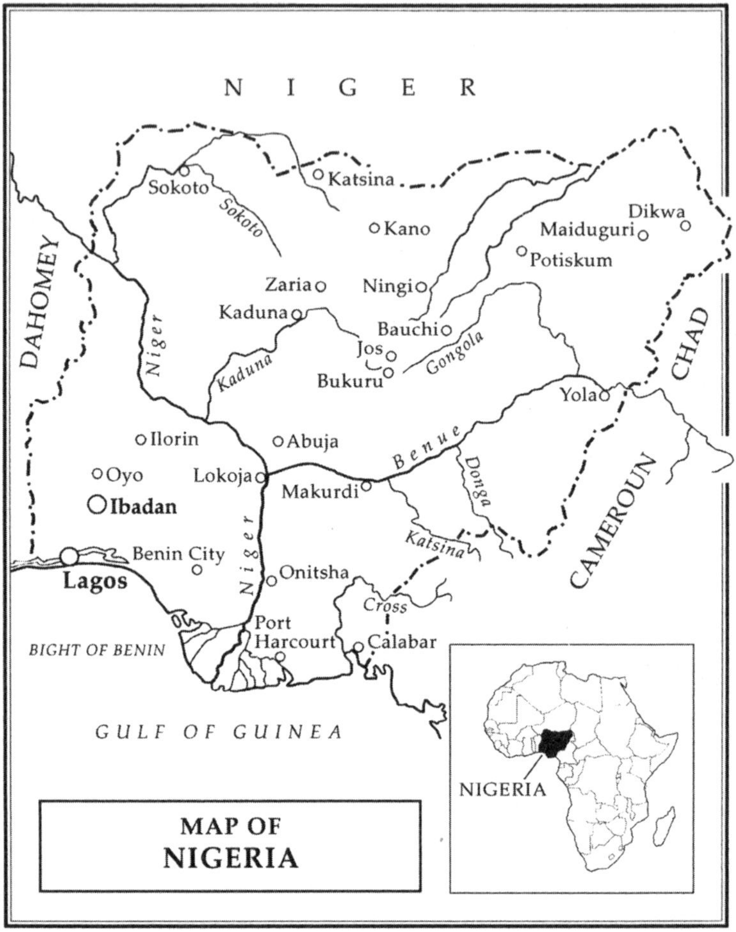
NIGER
DAHOMEY
CHAD
CAMEROUN
Sokoto
Sokoto
Katsina
Kano
Dikwa
Maiduguri
Potiskum
Zaria
Ningi
Kaduna
Bauchi
Niger
Jos
Gongola
Kaduna
Bukuru
Yola
Ilorin
Abuja
Benue
Oyo
Lokoja
Donga
Makurdi
Ibadan
Katsina
Benin City
Lagos
Niger
Onitsha
Cross
Port Harcourt
Calabar
BIGHT OF BENIN
GULF OF GUINEA
NIGERIA
MAP OF NIGERIA

Muxima
Mumbando
Malanje
Cuanza
Dondo
Calulo
Nova Gaia
Quitapa
Longa
Luime
Porto Amboim
Quibala
Quirima
Sanga
Novo Redondo
Vª Nado Seles
Chiengue
Mungo
Sautar
Cubal
Luando
Bimbe
Cuanza
Egito
Balombo
Bailundo
Munhongo
Catumbela
Lobito
Balombo
Silva Porta
Benguela
Bocoio
Hanha
Nª Lisboa
Gando
Catenque
Cubal
Dombe
Grande
Coimbango
Alto Cuito
Galangue
Chingoroi
Caconda
Lucira
Chicomba
Vª da Ponte
Lola
Vª Serpa Pinto
Caitou
Dongo
Cuchi
Area of
main map
DETAIL MAP OF
ANGOLA
ANGOLA

1. Captain William J. Chads.

1

Nigeria: The Bisichi Tin Company

William signed a contract to work with the Bisichi Tin Company in Nigeria from 12 October 1927 to 1 April 1929. He left London by train for Liverpool and soon settled in the restaurant carriage, which was unable to cope with the demand for lunch. As William had some sandwiches to eat he gave up his seat to an American who had only just managed to catch the train. She seemed helpless in many ways and suffered a further upset when her escort failed to appear at Liverpool. William helped her with her baggage and also managed his own, on which he had to pay for considerable excess weight.

The *Accra*, an old mail boat, was due to sail two hours after the train's arrival and from the start all friends of passengers were ordered off the ship. After all that a frustrating breakdown delayed the departure until midnight. The ship was full to capacity. William had a cabin right up in the front, which he thought might have been uncomfortable in bad weather but the seas were calm throughout the voyage.

The weather remained cold until the day before arriving at Sierra Leone when the sudden rise in temperature made the passengers thirsty and the stewards at the bar were worked to death until the heat subsequently became less intense.

In the 'hot zone' William slept on deck each night in his deck chair. This necessitated going to bed rather late, after all the women had left the deck, and rising very early because of swabbing the decks. After a bath and dressing he was back on deck before 6.00 a.m., usually before any other passengers were up. One night was more interesting. After finding a secluded corner on deck to spend the night, and just before dozing off, he noticed a couple nearby who appeared to be newly married. They were so interested in each other that the wife sat on William's lap before discovering he was there, whereupon they cleared off in haste and never spoke to him for the remainder of the voyage.

At first quite a number of passengers took vigorous exercise before breakfast, but the number dwindled rapidly. Inter-class cricket was played for a time, but had to be given up owing to the heat and because too many passengers disembarked at Sierra Leone and at Sekondi.

The American from the train distinguished herself by causing more scandal than one would have thought possible on such a short trip. Her capacity to find dark corners on deck late at night was limitless and, if sent below, she had a way of finding the wrong but fortunately always crowded cabins of friends where she would while away the next few hours of darkness.

William saw little of her except at meals when he sat at the same table. It was not unusual for her to miss all meals and then to appear on deck in time for dancing after dinner. Many people 'forgot' to say goodbye to her when she left, and her husband seemed in no way pleased to see her again when he came aboard to meet her.

One passenger on board was the president of the republic of Liberia who had been well received in the United States and had then gone on to France and England where further celebrations had been prepared for him. These powers were eager to obtain further rubber concessions, which he was in a position to provide.

At Sierra Leone, where the military detachment was dropped off, there were three divers in canoes. One, a Mr

2. Landing boats at Sekondi, Gold Coast (now Ghana).

Johnson, was considered to be quite an institution because he was always dressed in a loin cloth, stiff celluloid collar and bowler hat, which he used mainly for bailing out his boat. He normally dived for silver, but if trade was bad he tried to find copper. His blarney with the women on board was very skilful, especially with the younger passengers. The waters were shark free and there were divers for cash at other ports of call. Their method of rowing was interesting in that they sat at the side of their barges, facing ahead, and used a peculiar short three-pronged oar, somewhat like a big hand. They were noisy, happy lads and rowed in time to a chant.

On arrival at Lagos all baggage for up-country was transferred with the passengers to a lighter and taken across to Iddo wharf. The lighter had to circumvent a sandbank onto which it frequently ran aground. It took nine hours at Iddo to sort the baggage, clear customs and make arrangements for the two-day journey to Jos. The train was very crowded but the coaches were comfortable. Electric fans in some of the compartments worked efficiently, although William's did not. Sun-proof glass windows at all times protected the passengers in the carriages against sunstroke and made dull scenery look attractive. For the first day and a half it was difficult to get meals at moderate times – there were four sittings for breakfast the first morning, with the last starting at 11.30 a.m. Each sitting for a meal took at least an hour, and those who were in the last one for breakfast and the first for lunch had a practically continuous meal. Dinner in the evening was still being served at midnight.

The scenery consisted mostly of forests of rather small trees and little undergrowth; the stations were far apart and most of the towns and villages were small. The country was very flat until the train reached the Bauchi Plateau, at the foot of which William saw scantily-attired tribesmen for the first time. The incidental expenses of the train came to a minimum of a pound a day. Drinks were also very expensive, starting with beer at 1/6d (7½ new pence) a bottle.

The train arrived after dark; it had been eight hours late on the first morning and finally arrived three hours late, long before it was expected, so William had to wait an hour before the general manager turned up to greet him. He was then taken by car to Forum.

Forum camp

Forum camp was divided into two, one part for Europeans and the other for Nigerians. The former consisted of about ten adobe houses for offices and staff. These were built of sun-dried bricks and had cement floors for cleanliness and protection against white ants. Additional protection was obtained by tarring the bottom portions of the walls and whitewashing the remainder, which made it easier to detect them because once the ants were in the walls they were difficult to get rid of, the most usual remedy being paraffin or Flit. They were a serious menace and had an astonishing

3. William's first bedroom.

capacity to locate articles through walls. Within a few hours of some plans being put next to a cleanly whitewashed wall without any cracks, the ants would burrow through to come out at its point of contact with the wall and swarm all over the paper. Twice William arrived just in time to save the company's plans, from which the ants had already eaten nearly half an inch out of the margin. Most houses had to be reroofed every other year, partly because of the harm the ants had wrought and partly because of wind and rain damage during the wet season.

Cookhouses, servants' quarters and lavatories were all located far away from the European sleeping area. The camp was kept clean and a garden filled with fruit, flowers and vegetables seemed to grow well. Fruit consisted of papayas, pineapples, mangoes, a few limes and oranges. Papaya was available at all times of the year as a fruit or vegetable, depending on its state of ripeness. Vegetables were more difficult to grow successfully, partly because their seeds deteriorated and partly because grasshoppers

4. In Forum camp.

would attack the peas and beans just as they were almost ready for picking by biting through the stalks near the ground. There was no protection against them and they would fly over the houses for 40 to 50 yards at a stretch.

The local village was supposed to be laid out as a square of ten huts in ten rows. In fact there were nearly 300 huts and few Europeans could find their way through the village to the market unless they had been escorted there at least half a dozen times. Although it was swept every morning, the market place was always dirty because of the constant presence of cattle and humans. Slaughtering and cutting up the meat was done half a mile from the village. When the meat arrived at the market beating drums would tell everyone that there was fresh meat available for sale, usually at 8d (1.66 new pence) a pound. Shops selling alcohol were forbidden but they did exist.

Some headmen, older residents and traders would construct their own round huts with grass roofs, and each would build a plaited grass wall around his own group of houses. Some of these compounds consisted of five or six huts, thus allowing a separate one for each wife and one for the older children and cattle, goats or other stock.

Grass, maize and other crops had crept into spaces in the village, the outlines of which had become irregular through additional buildings; tracks would wind about between the huts leaving space free for the housewives to dry their herbs of many sorts in front of their doorways.

A motor road, which in dry weather was fairly good, connected the camp to Jos. The camber of the roads was excessive and at bends invariably sloped the wrong way; also, because they were made of laterite, the roads became slippery in wet weather. Ruts formed that could be more than a foot deep and ant heaps frequently appeared along the road, their size being limited to the clearance of the last vehicle to pass by. They were particularly dangerous to motorcycles.

The nearest railway station at Bukuru was roughly half way to Jos, but there were very few stores there. The land was fully cultivated wherever possible for up to half a mile in all directions, and in some directions for several miles. The country generally was undulating, sloping easily down to the Forum River and various small tributaries, the highest local hills being outcrops of rock a couple of miles away. Beyond the river were higher hills with a couple of small dense woods at their feet. Five miles north, on the edge of the plateau, there were other larger woods. This was where the pagans lived and it was said to be unsafe for a Hausa to go there, especially towards dusk.

Outside the villages pagan farms were scattered about at intervals and covered a large proportion of the fertile ground. Many tracks about a foot wide connected the farms to the local villages. Their cattle, which grazed in grass that was frequently five or six feet high after the rains, would often be sent to the local vet camp for inoculation against rinderpest. Some of this grass was cut for roofing, some was stacked, but most of it had to be burnt.

Burning the grass was systematically done by the pagans, who gathered from miles around for the occasion, usually mounted on horses and bringing all their war weapons like spears, bows and arrows and shields. They also brought their male children and dogs (and sometimes their wives) and would form a big circle around the area to be burnt. Any game trying to escape would be killed, though there only seemed to be about one hare for every hundred hunters.

The tin workings or paddocks stretched for miles both up and down the sides of the Forum River. They had always been managed by ground sluicing, the simplest method of tin winning. A big leat or ditch took the water from the Forum River ten miles above the camp and, running at a gradient of eight feet a mile, gained sufficient height above the river bed to enable the labourers to run water both into the sluice boxes and to wash away the overload of sand and clay from above the tin-bearing sands, or to wash themselves. Picks, shovels and head pans were the essential equipment, while crowbars were needed for boulders and sluice boxes for sandy ground that did not allow the use of a ground race.

The labourers worked in small gangs of seven to thirty men, each under a headman known as a contractor. They never knew what the contract was, but each week the ground shifted by each gang was measured as accurately as possible and the headman paid at a rate that varied partly in proportion to the volume to be lifted (taking into consideration the nature of the ground, the thoroughness of cleaning up the bedrock and the distance head pans had to be carried), and partly on lines governed by common sense. If a gang did a fair week's work the intention was that the headman would receive 12/– (60 new pence) and 6/– per man, 5/– per woman and 3/– per child. For no apparent reason pagans were paid about 6d a week less, although they frequently did more work than the Hausas.

5. Stripping overburden with water at Mission Creek.

It was impossible to measure the volumes lifted accurately and quickly, especially if the surface was rough or the bottom potholed or rocky. Considerable licence had to be taken with the final calculations to ensure that all headmen had sufficient money to pay a fair wage, and not to allow the headmen to be very flush one week and have too little the next. This was liable to happen if the depth sunk on a plot of ground was over measured in the first week and then followed by an underestimate the following week, the work done being the apparent difference between the two measurements. In the first case the labourers would only get their bare wage, while in the latter the headman would take his own requirements and then divide the balance, if any, between the members of his gang.

Complaints of any sort from the labourers very rarely reached the Europeans, whom they considered far too intimidating to approach on such a matter. Instead they preferred to accept the headman's word that he had received no pay for them, whereupon they would leave the camp and hope for better luck at the next place into which they strayed. By this means a few bad headmen could easily get a camp a bad name in an area where all labour was frequently on the move, few staying for more than three or four weeks in any one camp, no matter how good it proved to be.

If a case of a bad headman came before William he would suspend the person in question for long periods, and usually send him before the local judge (*alkali*) who dealt with all cases brought before his court with more firmness than mercy, and seldom allowed a man, innocent or guilty, to pass through his hands without receiving a sentence of either a heavy fine or a beating.

Forum was the company headquarters, the other sections being Ninghi (20 miles away and the poorest section), Bisichi, (five miles away and worked mainly with monitors and elevators, and with a pipeline giving a big head of water power) and the 1060 areas near Du Stream camp (the big section where the future of the company lay and where European staff increased from two to fourteen during William's stay there). Runners maintained communication between the sections, or in urgent cases from headquarters a car or lorry was used. A quote had been given for installing telephone communications but the price was too prohibitive.

The European section at Forum was kept clean enough to discourage flies, but they were bad in the bush near cattle and in the Nigerian village. Mosquitoes and sandflies were relatively sparse and for much of the year it was possible to sleep without a mosquito net. After dark praying mantises (some over six inches long) and other insects were sometimes attracted to the lights and were hard to drive away. Spiders, especially a flat type, abounded and were favourably regarded as they were said to eat mosquitoes. A number of snakes were seen at various times, all said to be very dangerous, though most were fairly harmless. What-

6. The big shovel at work at camp 1060.

ever the snake was it was usually called a black or green mamba. They killed several of William's chickens and, although never seen in the paddocks, they were seen in the office and under the toilet seats.

There were many varieties of ants but the white ones were the only destructive type. They were worst at the beginning of the rains when they moved from the open country in the form of flying ants and, collecting in houses and shelters, shed their wings and dug themselves in – if they got the chance. Black ants were useful because they ate the white ants. It was not uncommon to see a column of black ants returning from a raid, each carrying one or two white ants to be stored or devoured at leisure. The only good point about the white ant was that it never bit humans, whereas the other varieties did.

Hornets tried to settle in most houses, their sting being most unpleasant. They were disposed of by Flit or burning them after dark while in their nests resting on their honey.

There were a number of chameleons about, which changed colour at will for disguise. They had a curious walk

7. Big and little shovel and Fondson tractor at camp 1060.

and often stopped in the middle of a stride with two legs in the air for a minute or so, after which they would resume walking. While stationary the only movement came from their eyes, which they could swivel round to see in any direction. Although they looked quite harmless, William saw one put up a good fight with a 30-inch snake for two or three minutes. The snake bolted into the grass on seeing William and, though the chameleon still had a lot of fight in it, it died a few minutes later from the snake bites. There was one chameleon that appeared to be colour-blind – on khaki it turned to a brilliant green and on books and coloured cloth it seldom turned to anything approaching the right colour.

Lizards of all sizes and shapes were to be seen everywhere, some with bright colours and others dull. They sounded funny at night when they ran and slid over the corrugated-iron roofs, a sound that many people found frightening.

Bird life was rather scanty, but bush partridge and guinea fowl were easy to obtain for food. Bustards, both greater and lesser, were around during the rainy season. Plover were shy and not good for eating but suitable for a change of diet. Ducks were very scarce, but just off the plateau there was an occasional flock of green pigeon, which was

8. Sluice boxes at camp 1060.

the tastiest bird there. Their flight was amazingly quick and two were needed for a good meal. Other birds were plentiful but useless for food. They varied from crown birds and birds rather like overgrown dodos to a form of humming bird. Some of the birds were pretty while others were attractive songbirds. Wherever the labour moved there were plenty of carrion crows to pick up any rotting food.

On one occasion the manager of a property shot one of these crows for food. It was duly served up in the general mess that evening to three men, but one discovered that he felt too unwell to have any dinner that night and the other two were not very hungry. The remains of the bird were ordered to be served up cold for breakfast the next morning. The staff declined to eat any more on one pretext or another, so the manager, having helped himself generously after tasting one mouthful, threw some to his dog, which sniffed it and walked out of the house. A stray cat came in a few moments later and made a start to eat it, but soon stopped and was promptly sick.

The nightjar was the most common night bird and could often be seen a couple of hours before sunset. At dusk they looked very pretty in their zigzag flight with two pennant streaming feathers hanging out and looking like two bats fluttering behind the parent bird. There were a few bats at Forum but they seldom entered the house.

Owls of all sizes were numerous, the short-eared being the most common. They were a nuisance in the office because they kept a larder of putrefying small birds and mice on top of the walls and below the roofs of the office block. One evening three owls were killed in the accountant's room after they had made a considerable mess. On another occasion an owl stank William out of his office, so he sent a couple of henchmen to shift the putrid mess and the bird. He managed to restart work on a plan before the workmen had completed the job. However, William's flagging interest in their work was violently revived when the bird alighted on him and, digging its claws well into his back, proceeded to show disrespect in a manner that necessitated him having hastily to put on a clean shirt. The owl then flew out into the bright sunlight where other birds heartily mobbed it.

William occasionally saw a duiker (small deer), another larger deer and once a pair of ant-bear. Leopards and hyenas were said to be around Lower Forum but William was never lucky enough to see any.

There was no provision for tennis or any other sport with the Bisichi Tin Company, so shooting was William's only recreation. He seldom had to buy meat, but ammunition for his shotgun was hard to keep in good condition or to replace. Loose rounds carried for a whole day in the bush deteriorated rapidly, and at the beginning of the second day would hardly bring down bush fowl at short range. Ammunition stored in sealed tins lasted well.

There was not much disease about. Smallpox, blackwater and malaria were the most usual, malaria bringing everyone down occasionally. William took his quinine regularly, but found the usual pellets of five grains too much, but splitting them in two worked satisfactorily. One of the vets was nearly incapacitated by quinine, so he finally split his pellets into four, obtaining sufficient protection from that amount. A remarkably high proportion of men were sent home as a result of unusual accidents or trifling ailments like a slight scratch going septic. A few white men suffered from sunstroke and smallpox but William felt the country was much maligned for its climate

The 'pagans'

The local people William liked most were at that time rather pejoratively referred to as pagans, being neither Christians nor Muslims but probably belonging to a polytheistic religion. He found the men more honest and industrious than the Hausas (who are Muslim), and their women maintained a higher standard of morality, but were regarded as little better than cattle. In their normal environment, away from any camps or the Hausa population, the men and women wore absolutely nothing. Nearer other people, however, the women wore a bunch of leaves in front and behind, while the men wore somewhat less. Local superstition prohibited the women from wearing more than this, for they believed that clothes made them sterile. In rain and cold winds the women would look on and shiver while their men covered themselves with blankets and grass waterproof capes.

Most of the pagans lived in clusters of huts in the hills. With a stone wall surrounding each cluster, these settlements looked rather like small castle ruins. Each cluster would house a complete family and all its possessions. When a family grew too big for its village it would break up and form a second village close by. For some hundreds of yards round each village palisades of cacti were planted to act as a defence against enemies and to make boundaries to fields for crops or to confine cattle. The juices from these cacti were poisonous to humans and cattle. For example, William still felt a scratch on one knee six months later. Another European nearly lost the sight of one eye as a result of contact with the juice years previously.

The days of slave raiding were kept vividly alive in their minds through oral history and they still tended to fear white men. In fact if one walked towards them when they were alone, the young women and children (and sometimes even the elderly) would run away. In groups they were less timid and during the dry season, when there was no work to do on their farms, they would come to the paddocks in large numbers in search of employment.

Law and order was maintained by the local *seraki*, or king, who was answerable to the district officer for good discipline in his area. They had their own police force, which was quite distinct from the Hausa section. The pagans would take no notice of the Hausa police, whereas they were most submissive to their own. William always found it necessary to arrange for pagan police to come to Forum at weekends to handle all their own cases of excessive drinking. There were always plenty of pagan men and women in the local market and so long as they were sober they never caused any trouble.

A poll tax of six shillings a year was paid to the government and presumably there were also some local taxes as well. To pay these taxes and to purchase farming instruments practically the whole pagan population worked for the mining companies for two or three weeks a year.

There was much communal labour as well as what each family cultivated for its own requirements. Men and women worked in separate gangs when farming, the men doing the hoeing and the women doing most of the carrying. If there was insufficient work for one party then some of them would work for the other gang. While at work their meals were taken in a similar manner. The women did all the cooking and each sex formed its own circle to eat – the women taking care that the men had enough and fetched water for them when required. In the paddocks their food consisted entirely of cooked herbs. Water (which was not always clean) was boiled in a head pan and various grasses and meats would be added slowly and boiled until they formed a slimy yellow/brown consistency, which they would then eat with great relish.

Meat was a luxury of which they were very fond, though they rarely killed their own cattle or horses, which were regarded as important sources of individual wealth – in fact a horse or cow, depending on its condition, was accorded almost as much value as a wife – a good horse being more valuable than a bad wife. In some places grave robbing was thought to be prevalent, with corpses regarded as acceptable food. Consequently, Hausas buried their dead secretly and many spent large amounts of money on the prayers of priests to ensure that their dead remained undisturbed. The pagans did not, however, eat their own dead.

The pagans were not bound by any religious constraints on the number of wives a man was allowed. In fact the local king was said to have 13 who lived in the usual compound, each wife in her own hut, and each only appeared before the king when specially sent for, resulting in some being neglected for four or five months at a time.

There was a great deal of intermarriage within families. Special marriage rites needed to be performed and at these ceremonies young women anticipating marriage were often handled quite roughly. Some were even driven up to the ceremony tied to the ends of long poles on the grounds that they were unfit to be touched. To counter the effects of inbreeding the men of one village would agree to change places with the men of another village for periods of two to seven days every few years – an arrangement they apparently found quite acceptable.

The pagans seemed to be very clean people. At the end of a day in the paddocks most of them would wash thoroughly before going back to their homes. There were, however, exceptions among some of the older men who might bear muck marks from one week to the next. William's interpreter, Bekere, noticed that a mark from a clod of earth left on a man's back on Monday was still visible the following Friday. When one of William's peers asked if a certain man was clean Bekere, who took such infringements very seriously, called the man up and proceeded to smell him carefully, his final pronouncement being, 'Sah, he no be clean. I smell him fit to hear.'

With the exception of young children, all the pagans walked about with long sticks that were useful for probing the bottoms of fords, though they did not necessarily take them to the paddocks. Loads were invariably carried on their heads, especially by the women. Their heads were amazingly hard compared with other parts of their bodies, judging from how little they seemed to notice a hit on the head with stones from a headman accusing them of slacking.

The tracks were narrow and only permitted people to walk in single file. The women walked in front of the men but whenever possible they would collect in groups. Smoking was confined almost entirely to elderly women.

In the paddocks the women carried their babies on their backs in a leather sling, with no other protection from the sun or rain. If it rained they would turn their backs to it, so the babies got a hard upbringing. The only work women were allowed to do consisted of carrying head pans of dirt. Using a pick and shovel was a man's work. William learned this the hard way when he tried to make some women carry a section of a sluice box. In an instant he was overwhelmed by a row of flapping bosoms, whose owners were violently protesting – in a dialect he could not understand – that it was bad juju. He got the gist of the message and fled in confusion, leaving the men to do the work.

The wife of a very powerful *seraki* was the only woman William ever saw doing a man's work. On a special occasion, when there was a shortage of men and boys, she was permitted to lift head pans after they had been filled from the ground to other women's heads. That was licence indeed and it was remarked on for days afterwards.

Young boys and girls never played together. Boys liked to hunt lizards among the rocks with spears, bows and arrows. When older they became fairly good, in fact better than the Hausas, at finding other game and they proved to be keen and useful guides when William was unable to find birds.

The young women liked to make up using red ochre on their faces instead of powder and lipstick. Streaks and coloured marks were beautifying. Both men and women painted their legs and bodies in ways that bore significance for their particular tribe. Many different tribal marks could be seen in the paddocks, some an adornment and others a disfigurement. One tribe went so far as to make their women look really ugly from scars on their faces and bodies, which had in former times reduced their value on the slave market. With the exception of two tribes, the pagans seemed to be singularly ineffective fighters with their hands.

Some pagans, either deliberately or by accident, tended to start farming only a short distance ahead of the paddocks and, unless they were told to move elsewhere as soon as they started, they would lodge claims for damages once the paddock embraced their farm. One implement they used was like an enlarged infantry entrenching tool and it was usually made from an old pick or shovel. The handle sloped at a 45-degree angle from the shovel side of the tool and they would use it for digging between their legs as they progressed backwards. Long furrows were dug in this way and the crop sewn on the banks of earth thrown up. If this method did not appeal to the agricultural department, it was at least the result of many years of experience out there, and the government had been unable to improve on it.

The Hausas

The Hausas were easily the largest group in the population and their language predominated from the plateau area down to Lagos. William observed that many bad things had been written about this tribe and they seemed to live up to their reputation. Although some of the Hausas were good people, he accused many of being dirty, immoral, untrustworthy, lazy and ugly. They were very superstitious, which brought them under the control of the *mallam* of a form of Islam in which certain pagan beliefs played a part.

They were ruled by their own chiefs under British supervision and had a large police force who wore flowing red robes that were easily seen from far away. They often brought delinquents to the judicial courts where the *alkali* dealt out severe punishments for corrective effect. Litigation was liable to cost far more than the benefits sought. For example, a debtor might get a heavy fine and a thrashing with a possible prison sentence, but the one who instigated the proceedings to recover his money was also liable to a fine or a thrashing, apparently on the grounds that he was sure to have committed some misdeeds in the past. If his past had been blameless, this was a warning to keep it so lest something worse should befall him. Vagrancy was a serious offence likely to receive a three-month prison sentence. A note to the *alkali* saying that there were too many people at William's work place with no visible means of earning a living would cause a prompt round-up, and any man not actually at work or trading was liable to arrest, including those who might be taking a rest for reasons of ill-health or paying a visit to a friend. A man under such conditions was lucky to get off with less than a beating.

Divorce was easy and cheap to obtain unless the case concerned an expensive wife when the charges to be paid by the co-respondent would include the cost of her outfit plus her cost price with an allowance for depreciation.

Most men supported only one wife, and very few reached the luxury of four. Each woman was allowed a hut to herself, and each got her separate 'chop' allowance of 2/– (ten new pence) a week for food. The wives often stayed with the husbands for a long time and if they did leave they parted as the best of friends. Bekere, William's interpreter referred to earlier, had a wife for 12 years before she decided to marry another man. He sent her some cash as a reward for her long and faithful service, and expressed his willingness to have her return to his house if she became tired of her new man.

In the event of sickness they looked after each other well, with the men trying to bury their dead in places where the pagans would not find them. Many died from smallpox and fever, and sometimes death was caused by appendicitis or stomach troubles during the sugarcane season. William saved a number of lives from the latter by frequent doses of castor oil and insisting on starvation until the oil took effect.

The Hausas had two big festivals, the *sala* and the little *sala*. Before the latter they fasted for a month, eating nothing during the hours when the sun was up, with their capacity to work naturally declining considerably during this period. They generally arranged to work on the Sundays preceding and succeeding the festival and, insofar as such a thing was possible for them, they saved money beforehand to have a good time during the festival. On *sala* day the company allowed them one day's holiday with pay. As each festival approached they were anxious to borrow as much money as the company would lend them without any form of security, but these were the only times when the headmen would come up *en masse* to try combined 'eloquence'. On one occasion William lent each headman some money. As soon as deductions were made from their weekly pay, as had been arranged at the time of the loan, they complained that they were getting insufficient money, and at the same time they made suitable deductions from the pay of their labourers to cover (with a safe margin) the total sum they had to repay. So long as this form of robbery was confined to one or two headmen it did not matter because the people concerned were soon known and the labourers would not work for them.

William fixed a minimum pay for gangs of workers who automatically dropped out of work – men who never saved a penny. A man needed two shillings a week out of his six shillings pay for food and before the following evening all the remainder would be spent on drink, women or gambling. The headmen advanced 'chop' money to most of their labourers at the beginning of each week, interest varying between sixpence and a shilling for the loan of two shillings, and for men who had worked for a month they would make advances up to a fortnight's pay, keeping the interest high enough and taking the man's blanket and portable property as security. Occasionally, the man was able to slip away without payment, but he rarely escaped the pursuit.

European clothing was highly prized. One headman in the paddocks derived long service from an old frock coat and bowler hat even though wearing the coat over a flowing gown made him look quite ridiculous. For another, one of William's boiled shirts with the sleeves cut off served as his sole article of attire for weeks. Eventually he washed it – and it needed washing badly – but the resulting limp front disappointed him greatly.

In times of trouble the Hausas turned to their *mallam*s, many of whom were charlatans who twisted every event to their own advantage. They were ready to sell their prayers for a night or longer to the first ready bidder. If a *mallam* had a run of success he would raise his fees for services rendered, all fees being payable in advance, and he was not above touting for custom. In cases where they had no success it appeared to be either professional etiquette or law that the money had to be returned.

On two occasions William sacked a number of headmen and they all went at once to *mallam*s to ask them to pray that his heart would be softened sufficiently to allow them to stay on. They paid from £1 to £4 according to their needs and the reputation of the *mallam* to whom they went. Those who had been the longest with the company and had made the greatest additions to their huts (and possibly had small crops growing near the village) had most to lose and therefore were expected to pay more for the prayers that would keep them in employment for a bit longer.

Shortly before William's arrival in Nigeria there had been an eclipse of a nearly full moon. Although the people had been told about the forthcoming event in advance, as soon as the moon began to be obscured a great wailing went up from the local village. A story immediately went round that the sun and moon were quarrelling, hence first the peculiar colour and then the disappearance of the moon. Until the disagreement was settled there could be no dawn, just a continuous night.

Money and gifts poured in to the *mallam*s to pray that such a calamity should be avoided, and there were shouts of joy when the first halo of the moon reappeared. It was not known how much money the *mallam*s received that night, but one was presented with four women, all said to be young and negotiable if the *mallam* felt inclined to dispose of them, their value being naturally increased because they would be supplied with potent charms to hang around their necks to protect them from ill-health. As soon as the eclipse was over they immediately predicted that numerous and terrible diseases would follow. Trade therefore remained excellent, with so many people buying spells to counteract the diseases that were predicted to ravage the camp.

Adamu, William's chief informant on this matter, was sceptical about the value of the charms and felt that the masses who bought them were foolish and superstitious, but shortly afterwards produced two excellent charms for which he had paid ten shillings each. Two weeks later there were two short but very heavy storms. The *mallam*s at once came to the fore to explain that the storms were a direct result of their prayers and had been sent to wash away all diseases that would otherwise have killed most of the population. Presents flowed into the *mallam*s' hands.

Adamu spent a lot of money on them. When his wife died he spent £3 on prayers to be said for the safety of her body after burial. He felt the money had been well spent because there was no sign that pagans had tampered with the ground round the grave. The *mallam* had stipulated that his wife should have a secret burial at night, which might have added to the value and effect of his prayers. Soon after this Adamu was short of cash (which was not unusual among his people), so again went to his successful *mallam*. For a further £3 he obtained guaranteed prayers for a successful petition to the company for a loan of £15, without security, and for a rise in pay to enable him to repay the loan without inconvenience, for he considered it unlikely that the company would advance him the money without expecting to get it back. Unaccountably, this time the prayers failed and he demanded his money back. He had to take the case to the *alkali* before he got back even a small part, the balance going in the equivalent of costs.

Shortly before William left Nigeria Adamu was again in trouble. He had tried hard to save money, for he was the man with the most influence in the village and wished to uphold his position properly. Unfortunately, each time he built up a fund, it mysteriously disappeared. He therefore killed three fat chickens and made 'plenty good chop' for his latest *mallam*, and for that he obtained a chit to wear round his neck. He knew that in future he would be able to save as much as he wished.

On investigating the disappearance of Adamu's savings, William found that prior to each loss he had had at least one drink and had also given a few drinks to some friends, some of whom he had neither met before nor seen since. Sometimes, when he had had plenty to drink and had given his friends as much as they could take, he also gave them a present of money. He was never certain whether he gave them two pence or two shillings. Each time his savings vanished he felt ill and his friends did not come near him for days afterwards. That was now in the past and with a new charm to protect him it would not happen again.

Two weeks later, on a Monday morning, Adamu failed to appear for work. William sent for him but he was too ill to come, saying he had malaria – the symptoms being a dreadful headache and stomach pains. William sent for him to come at once and he arrived looking an absolute wreck, very sick and palsied, and was most grateful for a large dose of calomel. He felt disinclined to check if his savings had disappeared, but thought they had. The drinks on this occasion had started with vermouth and went on by stages through beer and a local drink to end with diluted petrol. He considered petrol to be a strong drink, but it had the advantage of being cheap and was essentially the drink for a strong man. He was very ill for a couple of days and after that he tried saving most of his pay in the company safe under the care of the paymaster.

Adamu was more intelligent than many of his peers and was anxious to fathom the simplest of tricks, all of which he regarded as juju. The greatest wonder of all was to see two magnets at one moment attracting each other and the next moment having the opposite effect.

The day Adamu purchased a motorbike was a great personal triumph. It was about the oldest one on the plateau and had received worse treatment than most. It was delivered by lorry and payment was to be made in another month or so, keeping open the prospect that something might occur in the interim to avoid payment altogether. The fact that the bike did not go did not dent Adamu's pride, indeed it seemed to increase it. He bought some petrol from the company and enrolled some admiring labourers to give him a push start. While being pushed over a long and wearying distance, he sat in glory on the seat, carefully moving all the levers and occasionally putting on the brake to show how well the machine worked. The admirers faded away and, being in danger of having to push the machine back, he turned about and returned to Forum still under manpower, but sounding the horn impressively as he coasted down the last few yards into camp. He went for several similar runs before the admiration of others began to flag. The machine was put away and almost forgotten until its vendor was due to return to England and wanted his money. Adamu saw but one hope of avoiding payment – he went to the best *mallam* in the area and paid him for further prayers. The prayers were more successful than he had dared hope. He was told to send the bike back at once if he could not pay for it because there was another taker to whom it might be sold. It went on the first available lorry, perched on top of various other stores and with Adamu seated in the sidecar. To complete the picture all he needed was a slave to hold an umbrella over him.

There was seldom any difficulty finding a new servant because dozens of them met every train, each carrying excellent references from previous employers. Some of these references were apparently made out for several people, for the servants had a way of changing their names in a perfectly convincing manner. As their previous employers had usually left the country there was no way of telling whether a man had earned his bundle of references or had hired them until he got another job. William was strongly advised to hold onto all the references for the duration of the man's service. Most men in the mining camps kept a cook/man Friday and an assistant, paying them 12/6d (62½ new pence) and 5/6d (27½ new pence) respectively. A gardener was a luxury. Married male servants were preferred to single ones, for they were less liable to get into trouble or to leave without warning, and they were likely to be cleaner. The downside of this was that the employer could overhear the woman's screams when the couple quarrelled. Wilson, a fellow employee at Forum, tried to economize on his staff and as a result they stole more than a corresponding amount from him in clothing and articles such as field glasses. He also set up a wireless and when he listened in for the first time he picked up a thunderstorm, so quickly turned it off. The next day, in his absence, his servant tried it and, according to hearsay, it was a long time before the set was fit for use again.

William was lucky in that his servant was more honest than most, though at first he tried to help himself to stores, but they were locked up and the chances of him being found out were high. On each occasion he had to pay fully for what he had taken, so he became exactingly honest and saw that his assistant did so too. This man remained with William throughout his stay in Nigeria. He was careful to give William only boiled and filtered water to drink and he kept the kitchen and house moderately clean. He fell from grace for a time while his wife was away and he was apt to wander down to the pub and get drunk. That did not matter so long as it did not interfere with his work, but the third time it happened a fine of five shillings was imposed with a warning that next time it would be ten shillings. He lasted some weeks before falling from grace again, and this time was also guilty of fighting and causing a disturbance. The problem was a palaver over a woman and he went to the *alkali* who ordered him to have two dozen lashes. He was unable to lie down for several days afterwards.

The man regularly borrowed money and always had an excuse. He repaid without a murmur and only once suggested a short delay. Within six months of his marriage he borrowed £1 to celebrate the arrival of his first-born child. Six months later he borrowed another £1 on the same grounds. Each time the child died before his wife returned to him, conveniently waiting until he had repaid the previous loan and allowing another for funeral expenses.

This servant took great care of William's chickens. He fed them regularly and ensured that not too many disappeared at any one time, although the visit of a snake one night and a mad dog on another could not have been predicted. When warned that he would be fined if the birds failed to lay enough eggs, the eggs appeared regularly. Later, William reduced the number of birds. He rapidly ate the hens until only one old one remained, yet that bird, with faithfulness not often encountered, continued to lay two eggs a day. If other people complained that their birds did not lay that was their affair and certainly not William's. The manservant always called the birds to him by crying 'Puss, puss' to which they came running.

The Hiscocks, a couple at Forum, employed a servant who assumed that if Mrs Hiscock failed to bath one day, she should have a double quantity of water the next because 'Mama no take bath yesterday. I be sure she be plenty dirty today.' He asked for a pay rise when she bought a pair of parrots on account of the additional responsibility and extra work entailed in cleaning out the bottom of their cage. Generally, these menservants disliked receiving orders from European women who had to elicit the support of their husbands. A quick beating was often considered to have more effect than insults or fines.

Most skilled labourers, such as carpenters and blacksmiths, were nominal Christians who had been trained in mission schools. William disliked them on the grounds that they were 'unattractive' and invariably less trustworthy than their Muslim fellow countrymen and because their slightly greater knowledge made them 'objectionable'. He saw them as always seeking to foment trouble, claiming all Christian holidays and Sundays, yet never missing the old traditional ones and retaining all the superstitions of their forefathers.

A few more observations about the Hausas

William was struck by the quite callous treatment some Hausa labourers showed towards each other. On one occasion at Forum a flood prevented them from returning to camp. The flood water was very strong, but a few of the better swimmers amused themselves by repeatedly jumping into a long straight reach and floating down on the flood to a suitable spot where they came out and walked up again to repeat the exercise. Eventually, one failed to reappear and another was badly bruised when he was carried a little further than he had intended to go. Apparently, no one made the least effort to locate the lost man; all present merely treated the matter with a shrug of the shoulders. The remainder stopped bathing, but only, it seemed, because they were afraid of suffering the same fate.

Despite their reputed desire to defraud everyone they could, there was never any fear of a man dying from destitution. Being a beggar was almost as profitable out there as it was in parts of London. The wish to give alms seemed to appeal most strongly to those who robbed their labourers in the most flagrant manner, while those who got drunk were frequently known to give away all their remaining cash to the first person who would take it. There was never a shortage of takers.

Their greatest fear of all, however, seemed to be the hospital. Death in any form was considered preferable and this fear made it more difficult to treat people successfully than it should have been. Many patients would only be taken to hospital in the last stages of their disease when doctors had little or no chance of curing them. Despite this fear they had a great love for white men's medicines, which for the most part they took fairly indiscriminately. Bekere was often ill and was always anxious to get hold of medicines of any sort. He would mix these with traditional treatments and believed that his faith and the mixtures were what obtained the benefits.

Even when white men went to hospital they were seldom expected to return. When William came back from a stint in the hospital he was told that he was not expected back for a long time. As a result his chickens had laid eggs for only two days after he left, but resumed laying at full strength on his return.

In parts of the country executions were still carried out by using a sword. The condemned man would be taken to the market place and, with his hands tied behind his back, permitted to walk about there chatting to his friends. The executioner walked behind him, sword on shoulder, with full discretion to delay the blow as long as he saw fit. The final stroke might be made within a few minutes or after the man had been waiting for it for hours. It could happen while the man was moving from one place to another or while he was talking to someone. It was not considered bad manners to remove a man's head in the middle of a sentence. The executioner was, however, disgraced if he required more than one blow to perform his task. The two figures in the act were naturally the heroes of the town for the day, the centre of all talk, and anyone who survived an executioner's sword became a man of some importance. The executioner's post was highly exalted.

When William paid a visit to Bauchi there was trouble in official circles. Apparently, there had been no execution there for two years and, as a result, the executioner was losing his status, which he resented, and for that reason decided to go on strike. If no victim could be provided and with no definite minimum number guaranteed for future years, he was going to resign from his post. Without a trained replacement available who could command popular respect, the disappearance of the present incumbent from public life was likely to be followed by an outbreak of crimes of violence. William never heard the result of the strike, but it was hoped that a public-spirited man would offer his services in time to save the executioner's self-respect with another faultless swipe.

2
Forum Camp

William was to work at Forum camp from November 1927 until June 1928. For the first two days after arrival he had his meals at the general manager's house. From then on he had to fend for himself, getting all his stores from Jos. which was 22 miles away.

On the second day he set out on the company's horse to see the work. The river was about three feet deep and the horse walked across it without any hesitation. As soon as he was on level ground again he decided to canter, but by the time the horse managed his second stride he had collapsed and turned a complete somersault. William managed to get to his feet before the horse did and once the horse was up he had a good look at some old scars on its knees. William then considered it advisable to lead the horse for the rest of the day. For the future it looked as if he would have to walk or use a bicycle to be sure of getting anywhere. It was easy to find the paddocks because the stream of labourers walking there daily had worn smooth tracks that were nearly two feet wide.

At the end of the first week William took over the outdoor work. At first the sun was trying, especially as there was no shade in which to find a temporary respite. Sunburn soon occurred, but having taken precautions in time it was not too bad. Soon William became acclimatized and appreciated the evaporation from the heat and occasional breezes, which kept his clothing dry no matter how much he might have been perspiring.

William found it interesting to watch the local black labour and he noticed many marks of rank among the numerous different tribes. Women outnumbered men and there were more children than adults. The children, both boys and girls, and some as young as ten years old, carried half-filled head pans throughout the day in a way that filled one with admiration for them but that must have been very injurious to their health. At the end of William's first week in charge he got rid of all the youngest children, and afterwards threw periodic fits of anger whenever they reappeared.

There were two working paddocks. William's first week in charge as a newcomer sent a signal to all the headmen to come to the office to complain that they had been defrauded of their pay on the previous Saturday. William passed them on to his predecessor Hiscock, who, suspecting that the complaint would come, had put himself out to double check the figures. There was no more trouble with the men at Mission Creek, but those at Barko Creek became slack. Their pay suffered as a consequence, so they tried another unsuccessful complaint. They did not realize that with his field glasses William could see a headman lying down from a long way off. There were also several occasions when William passed their sentries, who were posted at strategic points to give timely warning of his approach. At the end of a month William had to close the Barko Creek paddock because it was uneconomic to work. It was not difficult to select which headmen had to go because he could not employ all of them at Mission Creek. Two of the selected men who had to leave had served the company for many years and at the last minute were given a reprieve with a warning that they would be the first to go if their work did not improve. Although very grateful, they never recovered their money from the *mallam*s.

9. A monitor at Mission Creek.

The labourers were easy to handle and never thought of questioning a European's order. They were, however, easily amused. If one dropped a head pan, the rest would fall into ecstasies of delight and howls of '*waca*, *waca*' (clumsy) would go on for minutes afterwards. They would shout with joy irrespective of whether or not they had seen the pan drop. The small children were likely to drop their head pans into the water, thus causing a mix-up between the wash, the workings, the head pan and the child – all in the same splash of water, probably with another labourer tipping his wash onto the child. Tears were not uncommon.

The interpreter would take charge of the paddocks whenever the Europeans were not present. He worked for the company's benefit and was, as a result, unpopular with his fellow labourers, whom he loathed equally cordially and swindled if he got a chance. His English was not quite 'as she is spoken' in England and it took several days before William could understand him, and even longer before he could understand William. He knew the job well and was more reliable than some of the men who had been senior to him in recent years. A few headmen could speak some words that were considered to be English.

William's first Christmas Day fell on a Sunday. He had little to do and did not go over to the other camps with the result that the following morning he was the only person in Forum ready for work in the morning. He had to break into the offices because the keys were still out of camp and there was quite a row when Nicholls (the boss) turned up to find that nobody was around. He did not recognize Boxing Day as a holiday.

In the paddocks William was continually moving about on rocks, sand, water and stones. This, combined with other factors like the heat, quickly caused the stitching of his shoes and boots to rot. They wore out at an exceptionally rapid rate and before the end of the year he was beginning to feel anxious about his footwear. Other than that William on the whole felt well at Forum apart from two mild attacks of fever and one very violent cold that hung on for several days. Feeling healthy he ran much of the way out to the paddocks in the early mornings and in the evening a number of others enjoyed trying to race him back, occasionally getting there first.

In the New Year William was asked to take over the prospecting of the leases at the Du Stream camp. Hiscock had gone on leave and William had no one to fall back on in the event of need. The interpreter went sick the week before Hiscock left and, the following week, although nearly in the middle of the dry season, two days of heavy rain flooded the paddocks. No preparation had been made for alternative work and for a few days William might have felt despondent had he not already received one pay rise with the assurance that he would get another at the end of his first six months with the company. Production had gone up slightly and remained up until late in February when water began to be short. William had to search for new grounds to work and found that when prospecting soft and shallow ground the samples began to come in at a fast rate. In fact, he was snowed under with work in every direction. For a while longer the prospecting ground was within reach, across country from Mission Creek, and as a new horse had been purchased William could keep in touch with the situation. This horse could negotiate ditches and rough country with moderate safety. The old horse was sold to the king of the market for £3 – considerably more than it was worth.

In addition to prospecting at Mission Creek, William was in charge of two other camps just off the plateau. A local man called Raphael actually ran them but needed a lot of supervision. He currently owed the company about £60. A year before he had been £200 in debt, with practically no assets for security. As soon as Hiscock left and William took charge, Raphael sent in a number of claims for work done for him and for which he had not been paid, such as reroofing or building new huts. However, when William went to inspect the work Raphael was unable to show him any convincing evidence of it, so finally had to accept that William could not be taken in by such tactics.

On William's first visit to Raphael's camp he saw a number of men lying sick with smallpox in his *barraque*, and his other camp, No. 3, was practically deserted because the people had fled on account of the same disease. It took a whole day to visit these two camps, the further one being situated more than eight miles away. The track to it was passable for the first four miles and for the final stretch, but the middle part was so bad that William had to carry his bicycle above his head. Once having arrived, it was comforting to find that the camps were rather pretty, set in bush country with plenty of trees around.

In January a lot of tin was being stolen in the district and for a while William suspected that it was also disappearing from his tin shed. The tin would be locked up at nights and when cleaned sufficiently it was weighed and locked in a huge box from which bags were filled for each shipment home. Twice running William found he was two bags short and was slow to discover that the weighing machine had been tampered with during the annual stocktaking when it had been removed from its place to weigh stores for the paymaster.

William was not going to be caught again and resorted to lying out in the grass at night to try to catch tin thieves. This was unpleasant work that made him feel nervous. As he waited in the darkness he would hear a suspicious sound and suddenly find a lizard running up the leg of his shorts or trousers, or even running over his back or head – and his first thought was 'snakes'. The most reassuring sound was the singing from the village at night. While there was plenty of noise everyone was happy and William expected no

10. Sluice boxes and tin shed at camp 1060.

trouble, but when the stealing started the singing and dancing stopped in fear of the white man's wrath.

By the end of January the prospecting got out of reach on horseback and William got hold of an old motorbike and sidecar. He did not take too kindly to it at first because he had not ridden one for several years. The empty sidecar made the machine unstable on the roads and the controls were on the reverse side to those of William's last machine, so he instinctively did the wrong thing each time. Once he got badly off the road and narrowly escaped falling into a number of old prospecting pits, each of which was about three feet wide and twenty feet deep. It was hard work pushing the bike back to the road again. After this William always carried ballast in the sidecar – either a passenger or a couple of bags of sand.

Middleton, the mechanic, was due to go on leave in February 1928. Before his departure his servant went round the camp borrowing as much money as he could from anyone who was foolish enough to lend him some. He had access to the wardrobe and anything that seemed useful he borrowed, selling some and keeping the remainder for his personal use – telling his employer that it was all in the wash. He quietly left two days before Middleton's departure date, taking with him another man's wife – an expensive one. A great wail went up from all who had been robbed and Middleton was pained to find to what extent his wardrobe had been depleted; and his store cupboard had been forced open and emptied as well. A number of men came round to claim his furniture, all of which had been sold and paid for several times over to men in the *barraque*. These men wanted either the furniture or their money back; meanwhile, the servant had got away and was not caught.

The following Sunday William's houseboy came to him in a wretched state. He went down on his knees, grovelled in the dust and finally announced that he wished to have a money palaver. His wife had gone to another village and refused to return to him. He wanted another wife and had found one locally for £4. As a result of the other servant's defalcation he had only £2.10s left and wanted William to finance the remainder of the venture. The last wife had cost him £9 and had been a dead loss, but this one was cheap at the price. He dared not borrow in the *barraque*, loans at the time being hard to float and the rates of interest very high. He also thought that a loan from William would not require any security or interest. An additional advantage of going to William was that another man in the village wanted the same woman and any rumour of competition for her would at once send up the price. He agreed to repay the loan at the rate of ten shillings a week. He got the loan, although William felt it was too akin to slave traffic for his liking. To make repayment a bit easier he had his pay raised slightly, which sent an instant signal to his assistant and many others in the camp to come round demanding similar increases.

A few weeks later his first wife returned and a fearful fight ensued. William was not brave enough to face the excited woman so he sent for the local police who, with difficulty, removed her for long enough for William to have a meal in peace. It took two more days to get her out of the district and it cost William's houseboy four shillings to obtain his divorce in that time – a shilling more than he had expected. The houseboy was disappointed at the way things turned out because he still had hopes of recovering some of his original purchase price from some co-respondent.

When the Hiscocks left William moved into their house, which was more comfortable than his own. His first few nights there were disturbed, for lizards, mice and other creatures spent the night sliding down the corrugated-iron roof below the thatching, while the shrinking of the iron kept up a continual creaking that made William jump out of bed looking for thieves in or near the premises. William's old house was taken over by another newcomer who was put onto additional prospecting work, doubling the number of samples that had to be processed daily.

Smallpox had been rampant throughout the district and had now broken out in Forum village. The general manager's houseboy was suspected of having it and was taken to an isolation camp William had built of grass huts a mile from the village. At the isolation unit there was one guard for every six patients. To stop the patients going down to the village to beg, an allowance of 1/9d each was made to pay for chop. This left a fair profit for the people supplying the food. The general manager's houseboy was reluctant to return to a normal local diet and demanded, as his right, that he should still be allowed to have his food from Nicholls's store cupboard, as he had previously. Nicholls was annoyed to hear that he had been feeding on his salmon, jam and all forms of tinned food for the past few months. The houseboy also wanted his wife to attend periodically – a request that was refused. He escaped from isolation three times and was returned on being recaptured, finally being bound up to prevent further attempts. This did not stop him and the fourth time he got away without being recaptured.

A panic ensued so a doctor was brought out from Jos to vaccinate everyone. The Europeans were vaccinated first, except for William who was out so he managed to escape the ordeal. The servants came next and then the remainder of the labour force, who were told to roll up *en masse* for treatment, but only two came. After attempts to entice them had failed it was suggested that force should be used, whereupon all the labourers started to pack their bags and clear out of the camp, for they considered smallpox preferable to vaccination, which they saw as bad juju to be avoided at all costs.

The disease remained in the camp until the early rains. In addition to chop money and the 5/6d a week (27½ new pence) for each sick attendant's wages, the sick camp had to be maintained and whenever the numbers of sick diminished, those remaining would burn down the free huts, for which they could not be punished. The huts in the village where the sick had lived also had to be burnt to get rid of the infection. A fine of £1 was levied on any person who secretly kept a sick man without informing the office.

After repeated requests, the general manager visited the paddocks at the beginning of March and William was promised the necessary stores to put up a suspension bridge to carry a pipeline across Mission Creek. Work started almost immediately. The foundations for one side were on rock, while those for the other side had to be sunk to an unknown depth. Cutting into the banks of the creek a big pit was sunk 15 feet below the river bottom level, at which point seepage became too bad. Probing failed to touch the bedrock and loose sand was unthinkable as a foundation. Finally, 12-inch pipes were sunk into the sand to a depth of 20 feet, at which point they appeared to be well bedded and, on top of them, a seat of concrete was fixed. It took two days to sink each pipe because great care had to be taken while doing so because they were very fragile and old and driving them down with anything heavier than a 14-pound hammer caused them to buckle, after which they had to be lifted again.

There was then a long delay before anything else started to happen; it fact, it was not until the beginning of the rains in early May that the standards, cables and pipes finally arrived. Everything had to be carried out to the paddocks on the heads of the labourers. The uprights and cables were put up at once without difficulty, though the latter were not of the flexible type expected. Another delay followed owing to floods interrupting work in the paddocks and damaging the leat.

As soon as possible work on the bridge pressed forward. The first sling for the roadway was hooked on and clipped into position; the second one was prepared and ready to be passed out beyond the first when a worker stepped onto the first. It swung at the same time as the cables sagged and the man was lucky to hit the bank as he fell to the riverbed. He was not hurt, but after this incident the rest of the labourers viewed this work as bad juju and William had to fix each sling in position himself. Most of them were hauled up from below, with enthusiastic workers forever starting to heave at the wrong time and sometimes starting a tug-of-war over the cables. Several times William was nearly brought down because of this. When all the slings were ready it was found that there was insufficient rope to fix them rigidly while the pipes were put in position. By the end of the second day from the restart William had got most of the heavy 18-inch pipes in position, properly joined up, with only adjustments to the height of the slings and securing the inlet and outlet remaining to be done. He was dreadfully stiff after this unaccustomed form of labour, which tested his strength to its limits.

The next few days were during the *sala*, but a severe storm came up that ripped off the roof of the paymaster's house, as well as seriously damaging the office and a store. The river flooded and the rush of water severely damaged the main leat. The watchman then displayed some unnecessary initiative by opening the bypass onto the channel leading to the bridge. The unfinished bypass there was carried away and before the end of the holiday much of the end of the leat had washed out completely.

At Du the damage was worse. None of the narrow piers supporting a small pipeline across a dam had foundations. When the earth became sodden and the piers sank into it, the pipe bent as it sank and sprang a leak, which then cut off the power on the back side of the dam. It all happened at night and by the time morning came the dam was so weakened that it broke. It was the lowest of a set of three dams and took the strain of the lower half of the one above it. With the support gone, the second dam also broke and threw extra strain on the top one. That stood up to its work and did not break, yet the two lower ones had cost more than £1000 in labour alone to construct and now hardly a sod of earth remained to show for it. An estimated two months' supply of water storage had gone and it was expected that the cost of production would be materially increased. In the deep paddock 259 the water was flush with the top, and it took two weeks' continuous work before the elevators could empty it sufficiently for work to be resumed. Stories came in from other camps that the damage received elsewhere was more serious, some being crippled for the following season.

When work resumed William sent an urgent message asking for more pipes, which were produced reluctantly. No bends were obtainable and a sharp one was required that eventually had to be made out of concrete. The work was badly done but with continual supervision it worked out in the end. The intake for the pipes was too badly damaged to be repaired and a new one was made 100 yards away, from where the water was carried in pipes on piers to the bridge.

When everything was finished Nicholls came for a third visit.

Prior to Nicholls's visit to the site, a former company employee called Matthews had returned to Nigeria. Nicholls, who met him at Bukuru station but did not want him there, sent him straight to Du. He had arrived at Bukuru drunk and it was soon known that he had been drinking heavily on his journey. On receiving his orders he tried to drive the lorry on which he and his baggage were being sent and almost came to blows with its driver. A mechanic came to the driver's assistance and they arrived safely at Du, whereupon Matthews immediately informed Adams that he had come to take charge of the camp. A wordy argument ensued and Matthews duly passed from wrath to tears and pleading.

The following morning, Sunday, the two of them came to Forum to see Nicholls and settle the matter once and for all. Nicholls was conveniently out. Adams returned to Du and Matthews drove off to Bisichi to collect the kit he had left there. In a few moments he had replaced the driver and, nearing Bisichi, he decided to go and see another friend some miles further away. This part of the road was bad and, at the first bend, the lorry went into a ditch. Matthews arrived at Lower Bisichi on foot and in tears. After a few drinks and explanations about how the accident happened, he returned to Bisichi on foot – a distance of about six miles. He found the man in charge out, so commandeered a lorry there and, forgetting his baggage, drove back to Du. As seen from his tracks the next day, the lorry left the road about five times but somehow arrived at its destination; the next time it was loaded it promptly fell to pieces.

On Monday morning Matthews commandeered the company's best lorry to go to Jos, contrary to orders and before Nicholls could come over to talk to him. The driver's expostulations were ignored. Matthews returned that evening well stocked with provisions, including about 700 boxes of matches, a case of Worcester sauce, a gross of safety razor blades, two dozen toothbrushes and six cases of whisky. He also bought a couple of cars, but the owners refused to give delivery until they received cash. That evening Matthews proceeded to make two of the mechanics thoroughly tight so that they were unable to appear at work next day.

Matthews was then interviewed and told to go to Forum, where it was hoped he could do less harm. He was to be sent home on the next boat. He failed to arrive and disappeared into the blue, taking with him another lorry loaded with all his baggage and stores, as well as all the company's furniture other than the bed and a chair. For the next two days he floated round the country alone, dropping into white men's houses whenever he saw the chance of a drink. It is not known where he slept, but one couple were annoyed when he wandered into their bedroom while they were still in bed and, after vacantly staring round the room, floated out through another door. By the time the husband had put on his dressing gown and reached the front door Matthews had disappeared.

The police were put on his track but he did not appear again until the following Sunday afternoon when he turned up at Bisichi. He was promptly brought to Forum where he was let loose in William's house while his captor went to get instructions from Nicholls. In William's absence Matthews found an almost full bottle of whisky, which he consumed in less than ten minutes and he also drank most of a bottle of beer, the balance being spilled on the floor. He reduced the houseboys to terror when they were unable to give him any more whisky and then mistook William's bedroom for a lavatory. After that he was collected and taken to Jos where he was handed over to the district officer, who successfully lured him into hospital on the grounds that he had a badly sprained ankle. He was put to bed but escaped the next morning. Recaptured, he escaped yet again and was finally locked up in a safe place. There he refused to wear any clothes and behaved in a most unseemly manner. His baggage was found more or less intact, still in the lorry, which he ditched when another cross-country journey was attempted.

Matthews's main delusions in prison were that he was a successful racing motorist who had won many world records and that he had murdered his wife. This he justified by the fact that she had drunk 24 bottles of his best cherry brandy at one sitting – a deplorable act and worthy of death. In search of his wonder car he was taken on board ship to return to England under mild supervision. While seeking his cabin he heard that one had been reserved for a man who was insane, whereupon he lodged a protest with the captain and wrote to the company to complain of the indignity of having to travel with such a man on the ship.

When William returned to his house on the Sunday night he was enraged to find the filthy mess Matthews had made, so he took out his hunting crop to go after him. Nicholls tried to calm him down, but he was anxious to get both Matthews and his captor. It was fortunate that they did not meet because it turned out that Matthews had suffered a bad head wound during the war and that this, combined with excessive drinking, had caused his present trouble. The climate probably added to his problems, for he recovered normality very quickly after arriving back in England.

During *sala* William wanted to explore the area and obtained permission to sleep away from the camp. He decided to go to the Leri district and loaded his motorbike and sidecar with all the essentials for sleeping out, including guns, ammunition and special food and drink. The road to Ninghi was good and here he stopped briefly to see Fox who was unable to give him any advice on what was available to shoot, so he went on not knowing what to expect. The first major obstacle was getting across to the other side of the river. The river bed was dry and the banks at the side were steep and firm; at the third attempt he managed to climb the far side in bottom gear by having a short rush at it. The road from here on was said to be good in the dry season but now it was little better than passing over the furrows of a ploughed field after a hard frost. Occasional stretches of sand caused the engine to conk out even in low gear and, on one occasion, it took nearly an hour to get started again. The crank case was well in the sand and William could not attain sufficient speed to restart the engine. The kick-starter was out of action and a steep bank at either end proved too much to push up without assistance. Luckily, William found some pagans and, with their help, got going again.

The track lay through bush country and at times was clearly defined, but William had to stop every so often to remove boulders. In other places the track had broken away on one side or some obstacle necessitated a dismount and a careful push to the other side. The day's journey was tiring and made William and his engine very hot.

At Leri William heard that no game was known to exist in the district, contrary to what he had heard before starting. The nearest possible place was another 20 miles away along a bush track, but where one might have to wait weeks to get a shot. With this disappointing news William turned back for Ninghi and, with the added knowledge of the road, made much faster progress. It was an unsuccessful holiday but preferable to going with the others to the Jos races, where they had all had a bad day and most returned to their camps drunk. Another good reason to go away was that William had been told that beggars who wished to salute the white men and expected a reward for their trouble constantly pestered all those who stayed in the camp.

There were several storms in early May and William was caught up in a bad one while out shooting near the pagan hills. Being on the edge of a wood he took shelter from the driving rain behind a broad tree and, within a few minutes, he was standing with the water running above his ankles. As he held the rifle with the muzzle in the water a flash of lightning struck it out of his hand and he was thrown far into the air. After this William put the gun further away for the remainder of the storm. Half an hour later he shot a bush fowl from a distance of about 40 yards; he then consoled himself by thinking that his nerves were not too badly shaken after all.

Saturdays were always the most rushed days of the week. When William returned to his house late from the paddocks one Saturday in the middle of May, he heard that a mad dog had been killed after it had bitten two labourers and savaged a number of his chickens. William took the labourers to the vet camp to have their wounds cauterized – the vet was about to sit an exam to become a doctor. The vet was out and on returning to the camp William told Smith, the mechanic, about the event and he seemed to think it was rather a joke. Smith mentioned in passing that he had been nipped by a puppy that morning, but he knew the dog well. It was, of course, the mad dog and when he realized this he became most anxious to have everything possible done.

The vet returned far too late to do any good and suggested the dog's body be taken to Vom, his headquarters, where the pathologist should be able to test it for rabies. The body was put in a sack and William, with the mechanic, set out to Vom on his motorbike. Halfway there the tyres burst, one tube having a tear several inches long. Luckily, they only had to push for one mile, all uphill, to the nearest camp, from where they returned to their starting point. For the repeat journey they had a large lorry and completed the 16-mile trip just as the head pathologist was sitting down to dinner with guests. His wife was extremely annoyed that he had to leave the table at once to start work. He warned Smith that it might be more than a fortnight before any definite news could be given and strongly recommended that he returned home at once for the Pasteur treatment.

This was done, but there was no further information from his laboratory because, they subsequently heard, the pathologist was being sent home because while making up the serum from the dog's brain for injections into rabbits, the syringe burst and much of the liquid went into his eye. The vet claimed that he only delayed long enough to certify that the animal was mad before he hopped on the train to catch the same ship home as Smith. William saw the two black labourers who had been bitten nearly a year later and found them as well as they had ever been.

Du camp

June 1928 to March 1929

Hiscock returned to Forum at the end of June, shortly before his leave was due to expire, and Wilson arrived on the same boat to take over as general manager. For a few days there was uncertainty about the future but then, at very short notice, William was told to go and work at Du, and for the handover Hiscock did not even have time to walk round to see the work that had been done and the proposed work for the future.

On Friday afternoon William found time to run over to see his new quarters. Despite having been told that the place had been cleaned out and was ready for occupation, he found a single-roomed, filthy, dilapidated house two miles from the workplace without any furniture in it. The internal doors had disappeared, apart from one that was badly ant-eaten and the room was covered in bird droppings, broken nests and general filth. Added to this the house was situated on the wrong side of the river, which was very prone to flooding in the rainy season. The cookhouse was broken down and neither of the servants' huts was habitable, one having an ant-heap reaching into the roof but not visible from the outside. By this time both William's temper and his motorbike had broken down, but he was told that all would be rectified before Sunday and he walked back to Forum wearing new shoes.

Next morning William was still enraged and Wilson tried to be flippant, but quickly stopped. After breakfast William had recovered sufficiently to go round to see Nicholls, who had just sent a note for him. William said all that he felt like saying, after which, through careful avoidance like working the morning shift when Nicholls went in the afternoon and vice versa, he did not see Nicholls again for three weeks.

On Sunday morning William moved to Du at his leisure. The promises to have the place cleaned had been forgotten, no key could be found and William had to break into the house. Another hour was wasted trying to persuade labourers to carry his baggage across the river. It was late before the house was clean enough for dinner and bed that night.

William was told he would be needed for the early shift the next morning starting at six o'clock, but a storm and flood made him an hour late. On arrival he was told that the orders had been changed and he was not required until the afternoon; he remained working until ten that night. To start with, he was under instruction. Before that he had the pleasure of cleaning up after Adams's tailings and advising him on how to treat concentrates that were proving difficult to clean in any of the other sections. When William first looked at the boxes he was amazed, for no attention was being paid to the bottom end and the tin was shooting over it to be lost in a disgraceful manner. Adams and Acton would neither admit it nor permit a delay in feeding the boxes to get the losses picked up – quite rightly perhaps for it would have taken days to clean the place to William's satisfaction.

On the third day heavy rain fell early in the afternoon shift and there was a cold wind. Nearly all the labourers had left before dark and the few who remained were barely able to hold picks or shovels in their hands. William was shivering hard and the labourers must have felt even colder.

As soon as they finished cleaning up the first box William let them go home, but at that moment Adams came along and disapproved strongly of the weak course he had taken. Before they had finished talking, however, Adams admitted that the storm was exceptional and agreed to pay an extra day's bonus to those labourers who had remained until they were allowed to go home. About 4.7 inches of rain fell in three hours before dark and later there was plenty more. Most of the time the labourers were too blinded by the rain to be able to work and the rails were too greasy for the tractor to bring up the wash when the cars were loaded. All the men slunk off work from whatever point was not being watched, the two main points being half a mile apart, but it was now dark and the lights were not working well. As anticipated, William got back to find the river in flood and he had to wait some time before he could cross it to get home and have a hot meal.

The atmosphere after that evening was distinctly hostile. Every day numerous things needed putting right and they were all reported, but the hardest nut of all to crack was the loss of tin. Wilson alone accepted as positive evidence a collection of a quarter of a ton of crudes gathered from a patch barely ten yards square near the foot of the boxes. He tried to raise the question with the others but was told that it was only fine tin and as such it could not be caught. Nicholls soon learned to keep away from William and tried to prevent his daily reports being made to Wilson. Eventually, Nicholls lost his temper and indulged in a farewell burst. Neither listened to the other, but they both felt better after spilling the beans.

The steam shovels were continually breaking down and on days when they were being repaired William remained at paddock 259. There were also huge losses of tin here and William's first attempt to clean up one of the boxes took a whole day to clear and caused a lot more friction. The chances of regulating the water supply were remote, and it was nearly impossible to tell what tin was being lost because it went straight into the river and was swept away. To complete the clean-up operation another day plus overtime was required, after which the routine cleaning up in that paddock received the attention it should have had in the first place. At the end of the second day William was unable to get home because the river was in flood again – an event that was to become quite common.

At the end of July William took off a spare Saturday to go to Forum. He started the return journey on foot rather later than planned and forgot to take a lamp with him. As he reached Mission Creek darkness fell, with dense clouds rolling up and making nightfall very early. William hurried as fast as possible but the storm broke when he was far from a road or other suitable landmark and it was too dark even to see the outline of the hills. For a time lightning helped provide some light but when that ceased the heavy low clouds and pelting rain left him in complete darkness.

He was carrying his gun and an expensive textbook he had borrowed from Hiscock. To protect the book he wrapped it first in his Panama hat, then his shirt, and placed them in a haversack when the rain started and it was too dark to be seen by anyone. William headed for a road near the scene of old prospecting from Forum and, at about the time he expected, he heard running water ahead of him. The wind had lately been steady and had acted as an additional guide for direction, so he walked ahead confidently. Splash! He fell straight into the top end of the Forum leat. He got out on the opposite side feeling rather good, for now, being between the river and the leat, which were separated by barely 30 yards, he could not get lost. The road was a definite motor road with small ditches on each side. William went slowly forward towards it, probing the ground ahead of him at every pace. After what seemed to be hours of slow progress he came to water again, this time at the junction of the leat and the river and half a mile beyond the road, which he had crossed accidentally despite his efforts to find it. Since he was already soaked through, it made little difference to cross the river again in semi-flood. As he climbed the opposite bank for the third time at some impossible spot, a brace of bush fowl got up immediately in front of his face and caused him to fall backwards into the river again. He moved on and, once clear of the river, he saw the lights of paddock 259 about a mile ahead and tried to get through the intervening country. Before many minutes had passed he was in trouble again. He had slithered into some old workings and was anxious to get out anyhow and anywhere he could. When he reached normal ground level again he was still on the wrong side of the old paddock and, despite the encouraging lights ahead, William thought it best to turn back and try to find the road, which would be much safer. It took an hour to locate and then he nearly walked across it again, only recognizing it at the second ditch. During that time he almost fell into many old prospecting pits and was saved only by probing with his gun, which choked the muzzle with mud. From the road there was only a four-mile walk to Adams's house where he obtained a light and immediately returned home getting back to his house at 10.30 p.m. very ready for a bath, dinner and bed.

During the next week there was a lot more rain and the labour was at half the strength considered to be a working minimum. The rain continued and by evening the ground was so slippery that the tractors could not climb the greasy lines with full loads. William reduced the loads from seven to four boxes and all went well until someone in 'authority' appeared on the scene by paying a surprise visit late at night. He insisted that normal loads be resumed, but after half an hour of no progress being made, agreed to four being the maximum for that night.

With the heavy rains at their height, all the plant running itself in, and those in charge not accustomed to the work, nothing seemed to go right. It was usual for the lighting to break down to some degree every night. Silt at the intake of the pumping plant caused mysterious stoppages of the flow of water, followed by sudden rushes on resumption, making tin-winning harder than necessary. The headmen were exceedingly bad and the interpreters little better. There was no incentive in their contracts to make them or their labourers work, and the shortage of all labour, especially skilled labour, made it difficult to sack anyone as a lesson to the others.

One evening, when returning through paddock 259 to his house, William found the whole night shift asleep. No one was fined more than that one night's pay, though it was obvious that they had been sleeping most nights, but such was their discontent it was feared they would clear out all together if the punishment were any more severe.

The majority of labourers lived at Du and, having about a two-mile walk, they were apt to arrive half an hour late. After Nicholls's departure for England this was improved somewhat by more men being kept in huts near the work and many of the others being brought from Du by lorry.

On 19 September, as soon as Nicholls had gone, William heard that he was to return to Forum again. His feet had been very sore for some time and became worse after being in dirty water during the day's work. Eventually, the swellings were so bad he could no longer wear shoes and at this point he had to be taken to and from work by lorry and be carried over the river.

William's last night on night shift was a fitting culmination to all the previous ones. Labour had recently been improving slightly, but was very short again now and the weather was bad. The wheels of the tractor had worn badly and that night it became so dangerous that all trucks had to be manhandled up from the shovels to the boxes. Both the interpreters were away, genuinely sick, and every one of the arc lamps for the boxes failed to light. After wasting much time trying to get them to work, the man in charge went to

11. William's house at paddock 259.

get the chief mechanic who was entertaining a couple of friends. He arrived swearing, swore more when he heard about the tractor, and swore more than ever when one of the lamps exploded in his face. After that he retired to drown his sorrows and left William in darkness and having to deal with a hopeless task.

On Tuesday morning William was warned that he would be told to go to Forum on Friday. Later this was changed to some time on Wednesday. The lorry arrived at seven in the morning to collect his baggage. He had not started packing and, with his sore feet, it was a slow business. The first lorry left at eleven and did not return until two o'clock for the second load. Adams arrived just in time to see him leave and exchange a few parting sarcasms, followed by one or two personal remarks. William felt that Adams's face failed to express all the pleasure he should have felt at being relieved of his company.

The second lorry load contained a tin bath covered with wire netting and loaded with chickens. They disliked the mode of transport and, despite the heat in the sun, the hardness of the nest and the shaking they received *en route*, they managed to lay two eggs.

At Forum William could only have one of his previous houses, but with the promise of a further building being put up as soon as the dry weather permitted building work.

On Thursday morning William went to Jos to get his feet treated. The swelling had increased and for the next ten days he was confined to his room with one foot kept on a level with his head, which was especially unpleasant when sitting down to meals. An attempt to resume office work undid all the good the rest had done and at the end of a fortnight he had to be admitted to hospital as an inpatient. Treatment had to be drastic and on the first morning they removed much skin, a little flesh and a toenail in cold blood. While William writhed a nurse stood behind the doctor's back and giggled at him. William found this very annoying as he did not want to show what a funk he felt, so he confined his remarks to a foreign language, which was an unsatisfactory way of letting off steam.

After ten days in hospital he returned to Forum with instructions to do office work only for the next fortnight. William felt better for the rest as he was quite run down, and in the office he was employed on plan work. He wanted to do more work than there was available to do and finally Wilson dropped in with a textbook for him to read to occupy any spare time.

Things at Du were going from bad to worse. Breakdowns of all sorts were becoming more frequent, stories of local people being knocked about began to reach Wilson's ears and labour, especially skilled labour, became increasingly difficult to procure. There were plenty of carpenters and masons at Forum, but they were reluctant to go to Du.

By the end of October labour was plentiful again and was needed. The shovels were still breaking down regularly and the tractor petered out completely, so large numbers of workers had to be kept on to manhandle every truck up to the sluice boxes. The losses remained bad and were at last admitted to exist. With so much going wrong Wilson must have been hard put to think of any good news to cable home periodically.

In the last days of October it was announced that it was suspected that there were large tin deposits at No. 2 camp but that these had never been seriously prospected. A cable arrived to say that a gravel pump had been ordered for work there and that estimates for the cost of building a road to the site were to be sent in within a week. William's quiet office life came to an end as he and Hiscock spent a busy three days planning a possible route for the road and arranging for a gang of 50 men to clear away the scrub to permit a quick survey. By the sixth morning everything was ready with an estimate of £1500 for about eight miles of roadway. Nicholls had expected a figure of £300, but the price was raised because the work was being undertaken when the topsoil was baked hard instead of during the rainy season when the soil would be softer and easier to work. The next week they started prospecting pits for bulk samples and William was sent off to make a survey to see if it would be possible to construct a leat to supply water to the required area. A dry-weather road was begun to enable people to supervise the prospecting.

The Forum labour force suddenly leapt up to about a thousand, with accommodation in the local village stretched to its utmost. By early November it became necessary to subdivide the Forum command and William again resumed charge of the paddocks. Within a fortnight he had crashed from high favour to the other extreme. Though building had been in progress for nearly a month, William still failed to get permission to make the promised additions to his own accommodation.

Things were getting difficult because Wilson would neither go to the paddocks to see the work nor provide the essentials required to raise production economically. Pipes were urgently needed to prepare for water shortages, but no matter how the requests were made, in writing or in reports,

12. The gravel pump at paddock 259.

the bosses became irritated because the bank overdraft steadily grew.

They had not budgeted for the road construction, which lasted six weeks and cost £100 a week in wages, and labour expenses at all camps were steadily rising. The London office vetoed the road works, but the order did not arrive until after the work had been committed. New consulting engineers were causing Wilson concern and members of his staff at home were robbing him of his binoculars, clothing and food; one houseboy even took a new bed and mattress.

Late in November an edict was issued that again devolved responsibility for writing all the weekly reports on Hiscock. His optimistic approach had clearly been appreciated.

Early in December William started to work on leats for future river-bed work in camp 123. The work was very hard and, throughout that month, he averaged nearly 30 shots a day of blasting through rock. Prospecting at this camp and at Lower Forum (formerly called No. 2 camp) had promised great results, good tin showing in both places without too much overburden. While Forum was shining

London enquired why low production of poor grade tin continued at camp 1060.

Some 20-inch pipes arrived the week before Christmas, which was ample to meet the minimum requirements for working the paddock. One needed 15 labourers to carry each pipe to the site and a huge effort was made to have the water turned out of the river bed before the holiday began. This necessitated damming the Rein River and carrying its water along a double row of pipes for 400 feet before reaching one leat. The Forum River also had to be dammed high up, the water taken along a leat with a very slight gradient for half a mile and then carried in pipes across the river into a leat on the opposite side from where it was passed down below the bottom of the proposed work.

All the sods had to be carried a long way and the dams were very expensive. The trestle bridge had to carry a double row of the heavy pipes across the river, a width of about 80 feet, each trestle being about five feet wide and sunk three feet into the shifting sands. A foot of water was running at the time and timber or other makeshift rafts were unavailable to make strong foundations for the piers. It was hard work again sinking the deadmen under the uprights. The first attempt to turn the water into the leats on Christmas Eve was unsuccessful and required three or four days' work after the holiday to put it right. The Rein River caused anxiety throughout the season, for sand was being carried down into the leat where the reduced current caused it to be deposited and then it overflowed.

During the Christmas holiday William took a vast amount of exercise. On the first evening he went out shooting guinea fowl. He returned rather late but in time to see some people still arriving for the general manager's dinner, which he had declined to attend. When the camp 1060 crowd failed to appear at the time stated, cars were sent out to bring them over, for they were under the impression that they may have broken down *en route*.

13. Double 20-inch pipeline bridge at camp 25.

William had a good breakfast the following morning and set off at 6.30 for an all-day walk. He took the new road to Lower Forum where he crossed the Bisichi River, turned up to the Laffon River and circled back by the Ninghi dam to Lower Forum again. It was a good 40 miles and he got back to camp at 9.00 that night. He saw no guinea fowl but came home with a brace of green pigeon.

The following day he was too footsore to go anywhere but he did see a couple of ant bear in the distance. This was the only time he ever saw this rare animal and it was many days before he could find anyone to identify it from his description of its appearance.

Two weeks later, early in January, Hiscock was rushed in to take charge of camp 1060 and Fox came in to run the prospecting at Lower Forum. William was now back in favour.

The wash stacked round the sample pits in Lower Forum had given rise to great hopes and there was much rushing around pegging new prospecting leases to cover the whole area. Rumour reached the general manager and engineers

14. New road to No. 2 camp.

that other white men, apparently bent on the same job, had been seen in the district. In all haste the company's first applications were rushed to the resident in Jos, but the biggest and most important one belonged to the Bauchi province. William was told to take it to Jos and post it on to the Bauchi resident. William's suggestion that he take it to the Bauchi resident himself was considered a stroke of genius and was sure to be days quicker than trusting the post. His other suggestion that he take it on his motorbike was less highly thought of because it was still in bad working order and might have broken down on the way, and certainly would have failed to get him back without a night or so in the bush with a gun. He duly went off on the 100-mile journey to Bauchi by car. The road was not interesting, the first 40 miles beyond Jos being through fairly dense woods and the rest through light bush; for the last 60 miles there was no water to be seen.

In Bauchi William committed the atrocity of keeping a government office open for an hour overtime and then proceeded to waste another couple of hours with the official who had attended him. He got back to Forum at eleven that night to find Wilson still on tenterhooks about the result of the journey. After satisfying his first few questions William sat down to dinner in a business-like manner and was left alone to go to bed.

William's star was shining very brightly now. Wilson allowed himself to be dragged off several times to see part of the section at work and pressed (at last) to get out plans for the additions to his old house.

William had since moved across to Hiscock's house again, but thought it as well to look to the interests of his successors – he even suggested he stayed on extra time as he was due to leave in June. He still had plenty of outside work to do in connection with the new prospecting leases and, with careful rationing, he was able to supply many larders with birds without extra trouble to himself or wasting more than a few minutes of company time. Only one of William's proposals was turned down, entirely on account of the shortage of ready money to pay labour.

In the second week of February Fox came up from Lower Forum to take over from William before he sailed for England. He disliked all the walking William made him do when going round the section, and the way he passed on all the history he possibly could, much of which had never been put on paper. The new accountant was not in a position to worry him as he had his hands full in his fight against all at camp 1060 where he had created an uproar. Happily, William's last few days were at peace with all.

Packing for the final time looked most formidable until a number of offers were made to buy remaining food, lamps and other items, thereby reducing the volume considerably.

On the last night there was an attempt to hold a farewell party but, being in open revolt, camp 1060 declined to sit down to table with Happell. Forum dined alone and, after the newcomer was told not to talk 'shop', things went well.

The return journey

The next morning there was the usual last-minute rush and, after breakfasting with Wilson, a full carload came to Bukuru to see William off. Excess baggage was considerable and the late hour of arrival cancelled out any chance of reducing the surcharge by indulging in corrupt practices such as bribing railway officials.

William was glad to see the last of the Plateau. In the train he was fortunate to be taken for a 'political' and was able to reserve a compartment to himself, which was a blessing as the journey was very hot indeed. Much of the way he locked the door, stripped completely and, with the help of old newspapers, drew what breeze he could into the compartment.

At Iddo the train ran alongside the ship *Apapa*, which was to sail to England, but passengers had to wait from five until eight before being allowed on board for a bath and breakfast. Embarkation was very smooth and the ship was less than half full of passengers.

For the first evening William asked the second steward to put him as far as possible away from the other passengers and, as a result, he was placed three tables away from the nearest neighbour. As the ship proceeded along the coast fresh arrivals began to close in on William and he finally collected a young man, who had also come from the Plateau, as a companion. He found the women passengers uninteresting because they had all been on the coast for some time and looked very pasty.

At the islands they saw the first lot of 'home' complexions, including one who was more striking than the others and with whom William's new companion at once fell madly in love. She ran on an easy leash to an uncle and aunt, the only proviso being that she remain with at least two men at a time. The companion was quite affluent and he collected a very broke captain in the WAFs to make up his permissible trio. Ignoring all the other passengers, they conducted an ardent double wooing, but to everyone's surprise no engagement was announced at the end of the journey. The betting was all on the one with money, but the lady dispensed her smiles with wonderful equality and thus maintained the interest for all. The companion added to William's troubles by collecting the ship's flowers and bringing them to their table for port and liqueurs. William discouraged him as far as possible, but there was no stopping him once his beauty turned up. Nightly, the three would sit in front of William and he had to listen to the alternate proposals, occasionally receiving gentle taps on the shins from the lady who was thoroughly enjoying the situation.

After leaving Sierra Leone a plunge bath had been fitted up over one of the hatches, measuring about ten feet square and three to four feet in depth. Hardly anyone tried to dive into it but William's highest dive was from the promenade deck and only once out of a number of dives did he slightly hit his head on the bottom. Ladies were allotted the early mornings and, having tried it twice, decided it was too early. While the ladies bathed the captain stood on guard and gave William a long lecture on the habits of sharks. He was told they mother their young for a long time and that the young swim into their mother's mouth and belly whenever danger is about. He had opened captured sharks and found a number of these small ones inside, and still very much alive. When the danger passed the young would return to the sea again, and when they were old enough pilot fish would appear at the right moment to lead them off to fend for themselves.

Late one evening the other kind of shark invited William to play a variety of poker. One of the men warned him that he might have a very expensive evening ahead. However, by the end of it William had played double or quits with each man in turn and offered them their revenge whenever they wished to have it, but they never wanted it. That evening

William covered all his ship expenses and had a small surplus as well, feeling grateful for the instruction he had received in a South American mining camp.

There were the usual ship's sports, which were as dull as ever but helped to pass a couple of days more quickly than if nothing of the sort had been done. After a very smooth passage the *Apapa* arrived at Plymouth on 1 April 1929, a bank holiday, in very cold weather.

A summary of tin mining in Nigeria

Tin is a metal found in the form of the oxide casserite or tinstone. It can be found in other minerals like sulphides, such as stannine or tin pyrites, which are less important. Casseterite is obtained commercially from lode and alluvial deposits. The lodes are primary deposits that occur in rocks like granite and alluvial deposits are secondary, having been eroded from the lodes and washed downriver.

The younger granite rocks of the Jos Plateau were formed during the Jurassic period about 60 million years ago. Casseterite is found with rocks like biotite, granite and greisen. Most of the tin was alluvial but there was some lode mining in the region. Tin has been mixed in the Jos area since the sixteenth century and tin beads were found in the Nok Valley that were 2000 years old. In the 1880s an Englishman saw tin at the market in Kano and in the late nineteenth century Major MacDonald wrote: 'There is considerable trade done here in tin, which is collected by the natives in the streams which come down from the hills, they melt it down and bring it for barter in the shape of wire about half the thickness of one's little finger. The tin is of very good quality.'

In 1918 Roberts observed that the thriving Plateau tin industry during the nineteenth and twentieth centuries was confirmed by oral and archival evidence. It was manually done by men who worked the tin sands with water in calabashes. This caused the heavier tin ore to fall to the

15. Granite rock about 1200 feet high.

bottom of the calabash and it was then taken to Kano or Delma for smelting.

At the beginning of 1900 the British administered northern Nigeria and allowed the Royal Niger Company a half-share of the mineral rights in a large area north of the Niger and Benue rivers. Tin was becoming a valuable commodity for canning food, metal alloys and for the military. In 1902 the Royal Niger Company sent George Nicolaus to look for casseterite (black tin) near Badiko in Bauchi province. Here he found it was handled by butchers in the town. Most of the ore came from the banks of the Dilimi River near Tilden Fulani where it had been washed down from the granite hills south of the Dilimi River. By the end of 1902 the Royal Niger Company and a number of other prospecting companies took out mining licenses covering the plateau and by 1911 over 50 companies were interested in tin. By 1912 a peak of 82 companies were involved, spreading their activities to the lowlands south and southwest of the Jos Plateau in the Nok Valley.

Private operators managed the leases and over one-third of the total labour force was so-called 'tribute' labour. This was a freelance method of working in which either all or part of the lease was assigned to tribute labour and the tributer worked where he wished, either singly or with others from a small team. The tributer sold the ore he had gained to the operator at a fixed price per pound and lived in mining camps (*barakis*) built by the operator. In the camp the operator had some salaried staff like camp managers, headmen, storekeepers and watchmen. Salaried men were also employed to plan the working of the lease and to construct and maintain leats and sluice boxes. One result of this method was that the tributers worked the richer parts of the ground and quickly moved on to the next, leaving low-grade ground unworked and often covered with waste.

Mining activities today are very restricted. Alluvial mining has given way to deep-shaft excavations requiring heavy and expensive machinery that only larger companies can afford. Most of the small companies in and around Jos have folded up because of poor yield from the overworked river beds, but private operators are still working the small surface deposits. It is estimated that deep-shaft mining will continue for the next ten years. Some of the rivers mentioned above have dried up and are no longer mapped.

Because of the mining activities, old tools have been found in the region. Oldowan tools are the earliest stone ones and some similar to those found in East Africa were found at Pingell, 30 kilometres northeast of Jos. Acheulian tools have also been found on the Jos Plateau, particularly in the valley of the Forum River south of Jos. They have been found too in the Nok Valley west of Jos, which was also mined for alluvial deposits.

Nigeria today

During the nineteenth century the numerous factions of the Yoruba tribe were constantly at war in the southwest of the country. The people were moving from the open grasslands of north Yorubaland to the subtropical forests in the south. More aggravation followed with the influx of traders and the movement of freed slaves into the Lagos area. Various systems of government were tried and the warrior chief emerged. Under these men many people lost their personal freedom and were required to expand the agricultural base. The British controlled the Lagos area from about 1820 when they tried to eradicate the slave trade and encouraged numerous missionaries. They sought peace between the various Yoruba tribal factions and obtained treaties in 1886 and 1893. Gradually, British control expanded to cover a wider area.

Nigeria was divided into separate states, which over the years have merged, changed boundaries and subdivided. Although English is the main language in Nigeria today only about 50 per cent of the people actually speak it.

Since Shell discovered oil in Nigeria other mining activities have almost ceased as being less economic to run, although it is believed there is still some mineral wealth to be found.

Agriculture is a significant industry and Shell has played an important part in conducting research on disease resistant seeds that give high yields. It has also been instrumental in developing self-help agricultural communities to improve farming skills, marketing and setting up a cooperative organization. Rural areas are getting better roads, bridges and drainage. Special agricultural advisers are situated at the Imo, Rivers, Bendel and Cross River states.

There are 16 airports in the country, six game reserves and five national parks with a white-necked gorilla colony next to Cross River National Park. Northwest of Lagos is the Imeko Game Reserve, while east of Benin City is the Okhomu Game Reserve. There is an airport at Abuja adjacent to the Bobo Plains Game Reserve. Jos also has an airport and is surrounded by game reserves with the Falgore Lame-Burra Game Reserve to the north, Pamdam Game

Reserve to the south and Yankari National Park east near Bauchi. The best time to visit is between December and February. The animals are most likely to be seen in the early morning and evening and game wardens will take tourists out at 6.00 a.m. Hotel accommodation in the country districts is still very limited and the country still lends itself as an attraction to the explorer.

3
Angola: The Zambesia Exploring Company

William's next major trip abroad was on a year's contract with the Zambesia Exploring Company Limited in Angola. He got a visa for Portugal and Colonies dated 7 January 1930. His salary while travelling was at half rate, £11.6s.8d (£4.13s.4d a week) from 24 January to 9 February 1930. From 10 to 28 February he received a full salary of £27.3s.0d (£10 a week) and a ration and gun allowance of £10.7s.0d. For March and April his salary was £80 (£9.6s.8d a week) with a ration allowance of £31 and travelling expenses of £15.3s.3d. In May/June his salary and allowances amounted to £111 (£9.6s.3d a week), which was quite close to the £9.8s.0d that companies were paying secretaries in 1960.

24 January 1930 to 27 February 1931

On Friday 24 January 1930 William caught the boat train from Waterloo to Southampton. He left in something of a panic because going to the theatre to see *The Mikado* took precedence over packing. His parents went to see him off at Southampton where a representative from the office of Sir Robert Williams & Company met him and introduced him to Mr Frazier, his cabin mate on the *Watussi*.

Only half an hour was allowed for embarkation because, with forecasts of extremely bad weather, the ship sailed well before noon. Frazier was sick before they had even passed the Isle of Wight and many other passengers went to their bunks and failed to appear for breakfast the next morning. There was a lack of ventilation, which resulted in the bad smells that caused almost the toughest of the passengers to succumb to the high seas.

By Sunday morning the weather was slightly calmer for an hour or so and some of the sick passengers left their cabins to sit on deck to get a bit of fresh air. It was still considered too rough to take a bath but William got one by bribing the steward. At 11.00 a.m. a heavy squall struck the ship and sent it reeling over at a huge angle. All the chairs and tables hurtled across the deck, most of them were not fixed, but the fixtures of those that were broke and they too joined in the turmoil. Much of the furniture was broken and several passengers suffered fractured limbs. The noise of smashed crockery throughout the ship was terrific and added to this din was the chorus of screams coming from the terrified women on board. In the drawing room and smoking room the panels had been staved in. Many cabins were flooded and all the stewards' cabins in the stern had their portholes smashed in, as a result of which their belongings were either lost or damaged. One steward got washed overboard in his attempt to repair the damage and was not missed until it was too late for a rescue attempt.

After the squall came a strong gale and several ships in the vicinity sent out distress signals. The gale lasted almost until they reached Las Palmas, by which time some sort of order had been ensured by tying all the chairs and tables together in corners, lashed so that anyone who wished to sit on one had to have their feet up on another.

One evening William and three fellow passengers attempted to play a little bridge, but when all four, together with chairs and table, were hurled right across the smoking room, crashing into and seriously upsetting the captain and his friends, they decided that card games were too thrilling.

On the following Friday there was a meeting to select a committee for the ship's sports. Not many passengers came, but William (somewhat to his chagrin) was elected as the only British representative, the rest being Germans headed by a young prince. At the first meeting everything went well and an interpreter told William what the others were saying. It was difficult to get the second-class passengers to cooperate at first and to get entries for the many events, but these difficulties were overcome after patriotic appeals.

For three days all went well but by Tuesday war was breaking out on board. The Germans wanted everything their own way, the weather was hot and tempers were thin. The two big events of the day were the Ladies' Derby and a bridge tournament in the evening. From the start of the Ladies' Derby the Germans took charge and the Englishman elected to start the event was not even allowed on the platform. William, seeing he could do nothing useful and with probable friction looming, cleared out of the way. The Germans disqualified a Chilean woman under some pretext and in the final there was an appeal against the judge's ruling. The race had to be run again after the bookies running the totalizator had paid out; the extra event caught them without new tickets, so the astute reproduced the old tickets, leaving the bookies considerably out of pocket; the new winner of the final was not popular either. The gathering broke up with much grumbling.

When nearly all the English party had gone, the Hapsburg prince, the president of the sports committee and the purser appeared with a sack of clothing, each article of which was to be auctioned for the benefit of the seamen's missions. Almost every cabin had been raided without anyone's permission and these articles were now being auctioned. A record of sales was kept and, where possible, articles were knocked down to their rightful owners. William did not know about this until the sale was over and it made him furious.

The bridge tournament was due to start shortly after

dinner. It was a very hot evening and because of the heat William expected some withdrawals, but he did expect to see more than a bare dozen people appear. It transpired that the Germans had decided *en masse* to make their own celebrations in the only cool spot on board – the spot where the bridge was being played. Feelings were already very strained and to add to the stress the Germans became drunk and very noisy. William tried quietly to establish a more peaceful atmosphere, but with little effect. Then two of the bridge players tried roaring at the Germans and, within a second, there was almost a free fight between the Hapsburg prince and another Austrian whom the prince considered to be of low birth. William got the German tables moved a bit further along the deck and for a short time they were quiet, but on the next burst of singing and cheering the bridge players gave up in a fit of rage. William warned the prince that he would be 'for it' in the captain's office next morning and, having got that off his chest, he went to bed.

In the morning the captain had many visits before William got there to talk about the previous day's events. He confined his complaints to the raiding of the cabins. The captain was furious, for he could not countenance men raiding women's drawers and boxes without permission, and a complaint to his company's office with possible claims for damages would ruin his reputation. He also realized that it gave the steward an opportunity to help himself to any valuables left about and raised the possibility that passengers might claim that their valuables had disappeared. The prince and purser immediately stood to attention and before the captain had finished the prince was in tears. They agreed to make a public apology. William said that the English-speaking passengers would be satisfied with a private expression of regret and left the office for the captain to have a few final words with the prince and purser.

On leaving the office there was a deputation from the bridge players to see the captain to lodge another protest about the previous evening and the prince received a further battering. By the end of the morning he had got the message that his manners were not shown to the best advantage and that being president of the sports committee did not give him absolute power to do whatever he wished on the ship. The English party was distinctly unpopular after this and, being in the minority, kept a low profile for the time being.

A number of cubicles were erected on one side of the deck on the night of the fancy dress dance and it was understood that they were for general use. However, before dinner it became apparent that the Germans had reserved all the cubicles for themselves. The small English contingent then decided to take no part in the proceedings, though some had already changed into fancy dress. When the time came for those in fancy dress to parade it was lamentable to see some of the English party join in and march round. They secured no votes from the non-participating English and none from the Germans. The English tore up their voting papers and threw them overboard – an act that created much wrath against them. Soon the captain heard about their reason for not participating and then, but too late, the cubicles were thrown open for the English to use with profuse apologies that such an error could have occurred.

The following night was to be the last night on board for many, the night of the captain's dinner, prize-giving and revenge. Many speeches were to be made and all went well until after dinner when two or three people were unnecessarily rude to William who could not make out what was happening. When the time came for William to make his speech he sensed all was not well and was not going to run any risks until he knew what was on hand. It took him half an hour to find out that there had been a complaint about the accounts. He had vetted them and they were in order, but being written in German everything had to be explained and interpreted to him. The problem item was a sum of just over £5 spent on entertaining the committee of the third-class passengers for allowing first and second class to go to their dances, with no return visits. It had not entered William's mind to limit the extent of such entertainment to a specific sum of money and he was startled to find that they had consumed half a dozen bottles of champagne.

The English were satisfied with the explanation given, but the Germans had written it down as the committee's expenses and some know-all asked if that meant the money had been spent on the committee's drinks at meetings. There was fury all round and again the captain was called in, but still they could not control their tempers and when someone tried thumping the table to make his point the captain also lost his cool. They were told in no uncertain terms to behave or be escorted off the ship. In the end, the seamen's mission benefited by about £30, with most of the money being in the form of vouchers to purchase goods to a certain value at the barber's shop.

For William it was not a happy trip. Bridge went badly and it was rather expensive on board, but to his credit it was generally considered that he had been quite successful with the various brides on board whom he entertained while their husbands lay sick in their cabins.

Mrs Pittard and her brother were interesting passengers on board who had been to Las Palmas to collect specimens and bugs for the South Kensington Natural History Museum. One bug made its presence felt while they were inspecting some cattle and was duly collected from some obscure part of her clothing or anatomy and bottled. It is customary when submitting insects to the museum to state, in detail if possible, from what source the parasite obtains its nourishment. Her brother took this opportunity to make the most of this find. To their surprise it turned out to be one of an entirely unknown species and was therefore named after the person on whom it had been found. The authorities then wrote to say that they were willing to pay a reasonable price if she would spare them a few more samples. Further information was required on whether special precautions had to be observed when bathing during their breeding season – a question that hurt rather for she had had to do without a bath for 16 days just prior to the important capture.

After a 16-day voyage the ship arrived in Lobito, Angola, in the late afternoon of Sunday 9 February and William was delighted to move away from that lot of trouble. However, he found himself in a new lot, which again could not have been anticipated.

Mills, the new boss, came on board to meet William and Frazier. Their first impressions were that he tried to avoid paying for drinks but that he got on with the job to be done. In fact they were told they had to accompany him to the bush early the following morning. They had no time to sort out what food and kit to take, or to store unwanted kit, and were told that none of the party had yet been able to get their firearms out of the hands of the police.

As the ship was about to sail Mills discovered that he had brought no money with him, so they went to the customs shed to recover their baggage, but since the officials decided that it was too late to inspect it they would have to return in the morning. They were allowed to take what they needed for the night and it was surprising how much they considered essential, including tobacco and a cake, all going through undeclared and unexamined. The rest of the baggage was left in a pile, the two biggest boxes having been badly damaged during unloading.

Lobito town was full and many passengers had to go on to Benguela before they could find any accommodation. Frazier slept in the cramped company office while William had to double up with Mills in his room. Having been told in London that sheets and pillow cases would be supplied William was more than a little upset to find they were not and soon found plenty more to complain about.

The first evening they dined with the newly arrived Mr and Mrs Pittard who also gave them breakfast and made it clear that that was it. William did without lunch and they all had a horrid, greasy, expensive dinner at a local hotel that night.

During the day they returned to the customs shed. All arms and ammunition had to be left there for transfer to the *materiales de guerre* from where, it was hoped, they would be recovered in time. However, those deposited last October had still not come to light. It also seemed that the charges for them would be considerably higher than the rates the Portuguese embassy in London had given them.

The customs were kind and on opening one of William's 'suspect' boxes the customs officer swiftly closed the lid and it went through duty free. It was later explained that that was due to the policy of the company, as represented by their head man in Lobito, a Colonel Greenwood VC, who kept a more or less free house for the whole town and the Portuguese did their best for the company and staff in return. William estimated that he had got away with £5 worth of imports.

In the afternoon William and Frazier went to Benguela for more passport photographs for the authorities. Benguela turned out to be a big, dull, rambling gathering of wooden houses and stores, all very shoddy and rather expensive, even if the goods on sale were of the best quality. They returned to Lobito in the late evening.

On Tuesday morning there was almost no breakfast again because Mills's houseboy had got drunk in preference to cooking a meal. Mrs Pittard came to the rescue again for lunch, after which they set out for their first taste of the bush in Angola.

Frazier went ahead by train to Benguela, taking with him much baggage, while Mills and William, with some workmen and huge quantities of luggage, travelled 'luxuriously' by lorry. Frazier collected more baggage in Benguela and left a man there in charge of what he left behind. The party then headed for MacCay's camp and for half an hour or so

16. William's first camp 20 kilometres from Benguela.

got lost. When they reached the furthest point to which the lorry could be driven they unloaded everything and sent the lorry back for the stuff left behind in Benguela. It was a further hour-and-a-half's walk to reach the camp, nicely situated on low ridges in a big valley in limestone country and surrounded by light thorn bush up to 10 to 12 feet high. The nearest water was an hour's walk away at the point where they had to ford the river to reach the camp, and the nearest neighbour was half an hour's walk away.

In the first week the only wildlife to be seen was one duiker (a small antelope about 23–26 inches in height), a couple of dik-dik (small antelope about 14 inches high with long noses, a rudimentary tail and a tuft of long hair on the crown of the head), the occasional snake, some old tracks of leopard and kudu (the smartest of the antelope family, the size of a 14-hand pony, with long spiral horns and ears ten inches long) and a fair number of small scorpions. The lack of water kept mosquitoes and many other insects away.

The camping outfit was excellent quality but much was badly chosen and some essentials were missing. The tent measured ten feet by eight inside and had openings at both ends. In damp or wet and misty weather the mosquito net

17. A prospector's camp outfit.

curtain proved useful because the mist condensed on it and the interior remained fairly dry and warm. Without it all metal (such as guns when they eventually arrived) rusted quickly and clothing was frequently soaking wet in the mornings. The outer fly to the tent was rather too close to the inner for convenience in bad weather, but the overlap at the bottom was sufficient to provide cover from sun and rain for most of the spare kit and food. Two of the party managed to get spare outer flies to their tents, forming an excellent veranda and making much more room and a pleasant and convenient place for work and leisure.

A table was supplied but William's own folding one came in useful when he had office work to complete and the other table was being set for a meal. William's tin bath was better than the camp variety and served as a valuable alternative for washing clothes in dry areas because the Lux washing powder transferred the green from the canvas camp bath onto all the clothing in a patchy and inartistic fashion.

Cutlery and table crockery were poor and life felt more comfortable when china plates and cups were available. William's water cooler proved invaluable because those the company supplied were too small for butter and the cooler was much more effective than the traditional wet sacking. The kitchen utensils also needed supplementing; most were too large and there were too few of them. William had to take charge of the 'chop' boxes from Fortnum & Mason. The quality of the contents was excellent but the man selecting the goods included plum puddings for the tropics, and half the bulk of food could have been purchased locally at a quarter the price. The only vegetables supplied were about half a dozen tins of peas, which were supposed to last a month with no variety. Dried vegetables would have been cheaper, lighter and less bulky to carry. Another problem was the packing because anything from three to five boxes had to be opened to get the essential ingredients for a single meal. There was no provision for day-to-day requirements, so from day one they had to sit down with a seven-pound tin of sugar gracing the dinner table because suitable containers were not readily available in the bush.

For the type of work on which they were engaged William's oil-compass was worth more than the expensive precision instrument the company supplied, which, at the end of a tiring day was almost useless on account of its sensitivity. A barometer for heights would have been useful but an old hand clinometer sufficed. The greatest blessing by way of a work tool was a tropical umbrella, which saved many a drop of perspiration in the heat of the day.

The daily work involved walking considerable distances and Mills set the pace on the first two half days but could not sustain the pace he had set. The third day was the first full day and William took the lead early on and stayed there for the rest of the day going over very rough country. The next day Mills sent William and Frazier out alone, saying that he had office work to do. In fact, he was exhausted and after a quick walk round the camp lasting all of a few minutes, he spent the rest of the day reading the latest novel to arrive from England.

The weather was humid and sticky as there were a few light showers most nights and the daytime sun inevitably brought awful sunburn problems again.

On Saturday Mills and William left the first camp and returned to Lobito. When they reached the lorry they found it was devoid of petrol and the driver, after using the bus for a number of joy rides, had bolted.

They reached Lobito moderately early and proposed spending Sunday the 16th there and taking a trip northward on Monday morning. William was concerned about his belongings in the two crushed boxes, for he had heard that the Lobito stores were very damp and all clothing and boots had to be taken out and aired about every two months. Fortunately, that did not happen to William's belongings.

During their absence from Lobito much had happened, including a rupture of relations with the customs officials caused by Greenwood's 15-year-old daughter's elopement with one of the clerks. According to rumour everything had been carried out in the correct manner, except that Papa had not given his consent. He was apparently out of order in certifying her age as 21 only a few days earlier to get her a driving licence. After this incident it was probable that in future the customs were likely to be less easy to deal with. The place was alive with rumours to the effect that all concerned in the matter would be slaughtered, that the colonel was in prison for contempt of law, and so on. The colonel was able to look after himself, but there was general upheaval in the small British community when he got to know that several of them knew about the proposed elopement days before it took place.

After trying to sort out their affairs in Lobito they were again invited to dine with the Pittards in the evening, after which Mills took William off to get a glimpse of the night life. This was in a marquee cum bungalow where they spent a few hours drinking a little at fancy prices and talked about the one great theme of the day – the elopement.

On Sunday, with some trepidation, William took a couple of bathes in the sea. The water was filthy, partly as a result of drift brought down by the Cavaco River after a recent storm and also from the muck thrown into the sea at the other end of town, all of which floated down to the European end where people bathed with complete disregard for sharks, which were frequently seen within a couple of hundred yards of the shore. However, all bathers refrained from tempting them too much by swimming far out of their depth. They reasoned there were plenty of fish in the sea for the sharks to eat without coming into dangerous proximity of humans and therefore they felt safe. Some children spent the day in the water and no one in the last ten years had been bitten by a shark. Meals were still a problem so they tried to get Mills's servants to cook for them because the food in the hotel was too disgusting and they could not keep on asking Mrs Pittard. It was important to try and get some established place where they could get a decent meal in Lobito and Mills thought of another man's house two miles away for which they could use the lorry. However, as the servants who did the catering had to go some way in the opposite direction, the meals were always several hours late.

A trip to the wild north

Monday 17 February was the start of a ten-day trip to the wild area of the northern end of the company's concession.

Mills's adviser described the country they were going into as being almost uninhabited, covered in large woods and forests and not the best place to go to without any firearms. William therefore thought that it was lucky he had smuggled his pistol through customs because that appeared to be the only weapon they had.

Breakfast was planned for just after 7.00 a.m. and half an hour later the party would depart. William waited two hours before breakfast was served when Mensa, the great factotum, appeared. As an excuse he produced a new chauffeur – one Mills had sacked a week earlier – to replace the one who had just bolted.

After breakfast they drove the lorry back to their accommodation for loading, but it ran out of petrol long before getting there and they walked the rest of the way, sending the staff to get petrol and bring the lorry as soon as possible. Mills then went off to buy food to last them the fortnight they expected to be away, for the agreement provided that William should be given all meals at a cost of ten shillings a day. On his return Mills took William for a final glass of beer while the men did the packing. Recognizing his doubts, Mills went on to say that they were experienced packers and there was no need to check what was taken as it had all been put out and there could be no mistake.

Mills and William left at 11.30 a.m. and drove fast, only stopping once or twice to pick up a few things from various stores on the way. They did not stop for lunch, but by 2.00 p.m. they had a burst tyre. It did not take too long to repair and they hurried forward to make an early camp at 4.30 p.m. at the site previously used by Mills when he first arrived in Angola. The day's journey took them over low ground in the sedimentary rocks with few hills, the largest being at the end of the day.

While in Lobito William and Mills spent some time trying to put the brakes of the lorry into reasonable working order, but they were still very inefficient and in the evening, with torches and in pelting rain, they had another go at repairing them, getting wet and filthy dirty in the process. Meanwhile, the 'experienced' staff pitched the tents on sand and by the time William went to inspect the guy ropes the pegs were almost out of the ground.

Dinner was not a success either and the cooking was done under extreme difficulties. Between the rain storms insects came out and when there was a slight breeze blowing the lights went out. It was then found that no filter had been packed and one important food box and some cutlery had been left behind. Mensa was duly blamed for this and he in turn told off a small child who was supposed to assist him and look after Mills's gear. Mensa was ordered to drive the lorry back to Lobito before dawn next morning to collect the missing items.

Tuesday 18 February

At 4.30 a.m. Mensa set off in the lorry for Lobito, returning just before noon. On his journey he had heard from the drivers of passing lorries that the country they were going to was very swampy and although the projected route was about 200 miles each way they should change it because many bridges had been washed away in the rains. The final word of warning was that in case of accidents they would have great difficulty finding any carriers to help get them out of ditches or breaks in the road.

The drive started with a steep climb of about 4000 feet on a moderately good, very steep and continuously twisting road. There were long stretches where it was impossible to pass a vehicle going in the opposite direction and if one heard one above it was advisable to stop at the first open spot to let the engine cool until the other vehicle had passed. It was often dangerous and hair-raising.

After the hills they travelled fast along dull granite country without stopping until they reached Bocoio at 3.00 p.m. There they met Black, another member of staff, who had just set out to walk all the way to Lobito to get instructions from Mills and to solve his worries. His problems were due mainly to poor management and were dealt with in a few words. He said his camp was about half a mile from the road so they decided to spend the night there, but it turned out to be several miles, most of which had to be walked in pelting rain. Carriers had to be turned out in the rain to collect the tents and personal kit, which arrived shortly after dark. The ground was almost a morass or marsh, and setting up the tents in the rain left everyone feeling very gloomy, their spirits being lifted slightly by a hot meal and a whisky and soda.

The camp was devoted entirely to a search for diamonds but they never saw Black at work and William concluded the reason for this was that there was no work to be seen.

Wednesday 19 February

They had intended to have an early start but the party left shortly after 9.00 a.m., by which time the tents had dried in the bright morning sunshine. Most of the track lay through grass 12 feet high. When they came to the stream they had crossed the previous night when it was only ankle deep, it was now too deep to cross. After a bit of searching they found another crossing, which was waist deep and through which everyone had to carry full loads; some had to be helped to prevent the rush of water carrying them away into a cataract. Mills and Black were carried across by six men each, while William kept near the head of the carriers and used his fly-whisk to keep off the flies, which were mostly the dreaded tsetse that cause sleeping sickness.

This was a lovely and very interesting day's run. The whole countryside was full of bush fowl of several varieties and many were so tame that they could almost be caught by hand. The end of the day was rewarded by the prettiest camp site William had ever encountered.

Thursday 20 February

After a good breakfast William and Mills actually managed

to leave before 8.00 a.m. The road ran through wide rolling downs with rugged granite hills in the background. The roads on high ground were in excellent order and travel was fast. Occasionally, they passed gangs of labourers at work and during these first three days of travel they must have seen between 200 and 300 such men. Throughout that time they saw fewer than 20 women and these were seen in the small villages (shown on the map as large towns) that usually consisted of half a dozen houses, if that, with their retinue of traditional huts. If the local store or farm of the village fell into disuse then the whole village ceased to exist and soon disappeared.

It was unusual for these villagers to see any traffic and they all came out to see what was going on. The road workers were also interested in the passing traffic, but for different reasons. Their head tax was due and if unpaid they had to do road repair work, but there might have been a local administrator travelling with the party and if he saw anyone slacking he would suffer severe pains and penalties.

The diverted route to avoid broken bridges took them some way east and to the north of the required route to the farming area of Bailundo. Here the route turned westwards, the population increased greatly and the women outnumbered the men.

The road did not go higher than 6000 feet and the highest hills were no more than 8000 feet. In the afternoon they arrived at the first village, Vila Novo de Seles. By this time they had run out of bread, country wine and petrol. They stopped for an hour to replenish supplies and heard rumours of gold and diamonds to the south. Off they went after them as fast as possible along a winding road through very rugged granite country, the whole area being covered with huge boulders and upright pinnacles of rock. Just before dark they reached a place about 22 miles from Seles on the granite–limestone contact where they later actually did encounter signs of diamonds in the shape of Kimberlite and sundry associated minerals. They soon selected a spot that pleased their artistic and business temperaments and settled there for a couple of days. Despite mosquitoes and sandfly it was a lovely camp with good water nearby and a little labour close at hand. The grass was seldom more than six to eight feet high and the countryside was only lightly wooded with small clusters of trees. There were many bush fowl and also leopard and reed buck. They had seen little of the forests and none of the vast swamps about which they had been warned; in fact the whole trip had been quite delightful. The only work they had done was to get out of the lorry occasionally, walk a little way off the road and chip at a few rocks, or to stop off at an official's house to report progress and ask where to find valuable deposits.

It was a hot day and William and Mills spent a strenuous morning panning the stream for gold, tin or diamonds. They recruited two local men, which was useful but all instructions had to be given by signs.

Mills sent William back to camp for a flour sieve and awaited his return. While William had been making camp the previous evening Mills had gone out for a short walk and from what he had learned while out walking he was able to describe a far easier and better route back to the camp. He said that by this route it was only about 50 yards through the grass to a road with a bridge over the stream they were panning. William had heard these stories before, but dutifully set off as instructed. First, he had to force his way through very dense grass well above the level of his head up and over a hill, then climb through brambles before reaching the road after an hour's struggle. By this time he was extremely hot and unimpressed with the new route, which was almost impassable at times. Being new to the country he had the additional problem of not knowing what snakes or other dangers to look out for, but after a further 15 minutes he was safely back at camp. Despite being in a sweat he did not stop to have a drink or to sponge down before returning to the panning site by a much quicker and simpler route, only to find that Mills had gone, despite advising he would wait for William there. William shouted for him and looked around but he was nowhere to be seen, so he followed the planned panning route but still never found him. On returning to camp William saw Mills sitting down nicely relaxed having had a bath and lunch. William was furious and matters were not improved when Mills suggested that William must be tired because he was new in the country.

In the afternoon they used the lorry to get from one spot along the road to another. Being off the main road they drove mostly along tracks. At one point they got stuck in a swamp but managed to get out of it without sending for help. Their interest was kept alive by the bridges, which were liable to collapse at any moment, and the thick black sands that had to be negotiated.

Saturday 22 February

They set off at 7.00 a.m. and the terrain remained interesting. In the late morning William washed some samples in a river said to be crawling with crocodiles. They worked in a type of thorn gazebo (they called it a *zareba*) built by the local people and obtained their water from the river by means of a tin on the end of a long pole, which was a slow and laborious method.

They moved on and stopped at a German store to ask where any minerals of value could be found in the neighbourhood. The German said he could tell them where to find nitrate like that in Chile and a wonderful medicinal spring for a consideration. William knew there could not be nitrate in a swamp area, but they followed the German's advice only to find a big vein of calcite, which they had examined that morning, and the wonderful spring was none other than the trickle of water from a rock where they had been washing some samples earlier, and from which they went on to the river as the spring was nearly dry. The weather turned threatening so they drove back to camp for lunch and called it a day.

Sunday 23 February
In the morning they returned to Seles because of a shortage of petrol and food. Mills thought it worth sending a gang to prospect the area more thoroughly and to establish the extent of the interesting country. Meanwhile, he would go towards the coast with William and follow up the granite/limestone country moving in a westerly direction.

After completing their shopping in Seles they moved on towards the coast and soon reached a steep descent with no means of changing direction until they reached the low, hot coastal belt and there they picked up the road to Novo Redondo where they hoped to replenish their supplies of wine and whisky. It was very hot and arid so lunch was delayed for several hours in the hope of finding a cooler place to eat. The place they found was just outside Novo Redondo and it seemed logical to camp on the beach and have some refreshing bathing, but the beach turned out to be the town's refuse heap and it stank abominably. Undeterred, they unpacked the lorry in the best place they could find. The whole local population came out in their tens to watch their manoeuvres so Mills decided they should pack up again and move out of town, although they had been unable to make a single purchase there.

The road out of Novo Redondo was very sandy and often the lorry had to be pushed. Periodically, the wheels sank up to the axles in sand when the track had to be built up again under the wheels before the lorry could be pushed out.

It was a further 25 miles before they came to the River Cubal, the first fresh water they had seen throughout the drive. They camped here just after dusk in a cloud of mosquitoes, which put paid to any thoughts of having a bath.

Monday 24 February
There had been talk of a coast road, which they thought

they were on until they found it came to a dead end at the river and there was no road on the other side; the ferry that crossed the river was only used by the owner of the land on the other side.

Mills was anxious to get back to Lobito, mainly because of the lorry's brakes. The choice was either to return by the same route, which was unsafe until it was repaired, or to go on the coast road, which was unfamiliar to Mills. They set off towards Novo Redondo again and just before arriving there found the correct turning for the coast road, experiencing more difficulties with the sand along the way. With the greatest of ease they lost the road again and found themselves on fearful tracks and in the course of the day they covered over 93 miles before they reached a spot that was only 15 miles upriver from where they had camped the previous evening. Part of this journey took them through 13 miles of wooded tsetse fly country on the most awful track imaginable. The combination of bad driving and bad brakes made this the most exhilarating drive of a lifetime. Several times the engine stalled on small hills and the lorry went backwards, rapidly gaining speed until it ended up in a ditch or the men in the back were able to stop it with a block of wood under the wheels. In time the men in the back got scared and acted more swiftly to block the wheels when they realized that the lorry was about to roll backwards. Their prompt action saved them from one or two possible nasty accidents.

As soon as they were clear of the woods and the tsetse fly it was a relief to have a break for lunch. There was a house in front and while lunch was being prepared Mills and William went to ask the owner about the road ahead. The owner was not pleased to see them but did offer a cup of coffee. Conversation was limited because French was the only common language and no one spoke it very well. The owner would not allow Mensa, the interpreter, to come indoors, but even without his help they understood that the road came to a dead end and they had to go all the way back through the wood again almost to Novo Redondo.

A heavy shower of rain had made the surface of the track even more greasy and dangerous than before, but fortunately Mills was too tired to drive and he let the chauffeur take over. Although his driving was better than Mills's he still managed to keep his passengers' interest alive by stalling the engine a few times in awkward places. The flies also continued to enter the lorry and were hard to dislodge. William's fly-whisk helped to keep them off but they moved far too quickly to be killed. William received one hard bite while the other passengers suffered many. Once they were clear of the wood again they were driving through strong overhanging shrubs and a branch caught Mills across the eye, which caused him to scream with pain and he was lucky not to have lost that eye.

Eventually, they came to a turning and branched off on another road, which they hoped might be the right one this time. The surface seemed to be even worse than what they had already gone through, if that were possible. The bridges were potholed and very broken down and it was almost a miracle that dangerous and daring driving got them safely to the other side of the bridges or culverts. After another belt of tsetse fly country they finally reached the River Cubal again. There was no bridge there but there was a ferry, though sorting this out was not a simple task either. First, the lorry had to be unloaded and by the time some local men had been collected to work the ferry two hours had elapsed. The ferry had to be brought over to the other side of the river during which time the towing rope broke and there was another long delay in finding a replacement.

The ferry operators did not want to take the lorry across the river that night and their awesome appearance made Mills and William think it best to stop the night on the opposite side of the river. Mills by this time was anxious to get back to civilization and would have pushed on much further that evening had it been possible. They got across the river by 9.00 p.m. by which time it was dark, but they were assisted by the lorry lights and a hand electric torch. That night they only pitched one tent, in the middle of the road, but as there was no traffic they were not disturbed.

After a good night's rest William remembered to adjust the film in his camera while it was still dark. Mills was conscious of the slight rustle of the paper and in the morning looked to see what had been stolen and asked William if he had heard someone enter the tent during the night. There were stories of cannibals in the district and several Portuguese had apparently been eaten. Many of the local men had filed teeth, which strongly suggested that they may be cannibals.

They were up early on a damp and drizzly morning. William had a bath with almost warm water from the river and by the time breakfast was ready the tsetse fly had disappeared. Mills was getting increasingly anxious not to spend another night on the road and was keen to get going, but the men were slow and it was 8.30 a.m. before they left. Mills then drove full speed ahead, averaging about 13 miles an hour over fearful roads covered in grasses and often heavily overhung by bushes. At times the route was so indistinct that one of the men had to get out of the lorry and walk ahead to mark the way. Many things called bridges were taken without even stopping to see if they existed, which resulted in many heavy bumps! It was the severest test any lorry could have, yet Mills was hurt when told he was driving recklessly. He slowed down a bit, but at each slightly open space he put his foot down and everything and everyone bumped about more and more violently. Luckily, there were no hills on which to stall the engine and after two-and-a-half hours they reached the road they had originally intended taking, which was falsely named the coast road. Next they came to the Uaque River, which was in flood and the bridge had been mostly washed away. It was

18. Flooded out from the river bed at camp 123.

impossible to cross so it looked as if they would have to spend another night under canvas while waiting for the water to fall and the bridge to be repaired. William took this opportunity to produce a typewriter, which annoyed Mills considerably. He decided they should do some prospecting and went for a short walk of about half an hour.

After lunch they had the most extraordinary good fortune. In a hut on the other side of the stream they found a number of heavy planks that had been left for rebuilding the bridge at the end of the rains. These were soon utilized and William used his army training to rebuild the bridge and in time they made their adventurous 'successful' crossing.

From this point on the road was good and the lorry was driven as fast as was reasonably possible. The next stop was for the ferry across the Bolombo River at Egito – a dirty little coastal one-horse town of about a dozen Portuguese houses and stores and a small indigenous village that is shown on maps today.

It was an hour before the crossing could be made and a further hour was spent shopping and getting more petrol.

After an hour's drive they camped near a well at a site Mills had used before. On that occasion there had been a fight between a donkey and a leopard right at the entrance to his tent. Again it was a filthy dirty place and difficult to get it clean enough for habitation before dark.

Wednesday 26 February
They were on the road early in the morning and travelled fast until 10.00 a.m. when they came to another river where there was no ferry and the bridge was completely washed away. The river was about ten yards wide and the rushing water about seven feet deep. Workmen had already started to repair the bridge but the job was going to take time. After much trouble they managed to get a boat to ferry them across with a minimum of baggage. Six men handled the boat and there were four of William's party in it along with their limited baggage. It was a deplorable exhibition of oarsmanship. Three of the rowers were sprawled partially into the water and the boat travelled mostly broadside onto the intended direction; and they were carried more than 100 yards downstream before being able to land.

Once on the other side it took two hours before they managed to find and hire another lorry to drive the final 31 miles back to Lobito. It cost only £1 and they arrived in the early afternoon. The tents, lorry and personal baggage were left with a guard of two men until the bridge was repaired.

Poor Mills, who was feeling ill, went straight to bed on arrival. Strangely, his indisposition allowed him to go on the tiles until well after midnight. He got up late the next morning and sent a man to tell William that he was being given another tent and was to catch the 1.30 p.m. train to join Frazier and Blackwell near San Pedro. William told him that this was impossible. The other two were trying to get their firearms back and William and Frazier also spent much time trying to recover their passports and other documents from the authorities. By the time the train had left William had had no breakfast or lunch and Mills seemed more pained than surprised that William was not on it as he had already sent out his tent and some stores with Blackwell. Mills said he wanted the office (William's bedroom) to write his report on this trip, but he wrote his notes in his bedroom and called William in from time to time to comment on what he had said. By evening William had packed and was ready to take the goods train after lunch the next day. He did not have time to get food in Lobito and took company 'chop' boxes containing a three-month supply of badly packed and poorly selected food.

4

The Blackwell Partnership

Many authorities had to be consulted before William was able to receive permission to travel on the goods train to San Pedro on 28 February and every care was taken to see that he did not leave the train until he reached his destination. It was a rather uninteresting journey of three-and-a-half hours covering 37 miles. On arrival at San Pedro station Frazier and Blackwell greeted William and they had his tent there ready for him.

28 February to 25 April 1930

After so much travel William might have been feeling rather tired, but he felt fit and well, only suffering a bit momentarily from the food served in the Lobito Hotel. He had a slight festering sore, which took some weeks to heal, and he snapped a tooth when biting the local bread. At the best of times the bread was hard, but after ten days it became rock hard. At first William's jaw felt very stiff from this unaccustomed work, but it gathered strength as time went on and the bread got harder. The following morning MacKay was due to arrive and most of the men from the two gangs of workers had been sent to help him with his move. It was therefore proposed that they should all move forward together along the general line of the railway.

Sunday 2 March
It was a day of relaxation and looking round the area of the

camp. While they were out they found a small boy had trapped a beautifully marked leopard by the railway line. When he was setting the trap the boy realized he had forgotten the usual bait and went back to collect a goat and on his return was very surprised to find the leopard already there waiting for him. Two of the party went to have lunch with the station master and engine driver and all returned late afternoon thoroughly drunk. They were still around when it was approaching dinner time and by now desperate measures were needed to get rid of them. They were finally taken most of the way back to their house and were too drunk to return and cause a further nuisance. After dinner William enjoyed the luxury of playing his gramophone but on returning to his tent he lost his one and only needle.

Monday 3 March
The day's journey was in open, light, flat thornbush country but the men only managed six miles because, although the distance was not great, water had to be carried and this was all they could manage.

Tuesday 4 March
Continuing along the railway line it was only about three miles before a waterhole in a rock was found. Here they collected enough water to cover their requirements for the next part of the journey. Gossip centred around Mills and William was surprised that none of the party had a good word to say about him, not even as much as being moderately efficient in geology, his speciality.

Wednesday 5 March
Frazier and MacKay left the rest of the gang and moved across to the Cavaco River about nine miles to the north.

Thursday 6 March
Mills sent a note recalling Frazier urgently as he was required to prospect the ground south of Vila Novo de Seles which they had cursorily been over very recently.

The camp was made near the railway line and on the way they spotted about half a dozen kudu, including two large bulls. To the south of the camp lay Mount Sohoa, the only high ground in the district and William walked to the top to obtain a good view of the plain for mapping purposes. On his way back he came across a hyena and brought it down with a shot from his pistol.

Friday 7 March

Things were getting complicated. Frazier suddenly arrived back at his camp because he had to wait another day before he could obtain permission to travel to Lobito by goods train. Blackwell accompanied him because he had to bring out more cash for the gangs, which should have been given to William a week earlier when he left Lobito. It was therefore decided that William should stay with the camp in its present situation as moving forward would involve Blackwell in a long walk in the dark from San Pedro station.

During this enforced delay disaster struck. They encountered the first white ants they had ever seen and they headed for William's chop box. He took this very seriously and took the necessary precautions to get rid of them. They were also out of whisky because a case forwarded on 27 February from Lobito had not arrived.

William spent the weekend planning the next camp and moving a few loads forward. The missing needle for the gramophone was worrying William and after some thought he remedied the situation by cutting some special thorns and burning hard enough points on them to produce quite reasonable results without damaging the records in any way (and these 78 records are still playable today, more than 75 years later).

MacKay reported seeing a considerable amount of fresh leopard spoor all round his camp, as well as other big game in the district. While Mills was taking Frazier up north William got in touch with Colonel Greenwood about getting his firearms back. The reply that came back said that the delay was because no duty had been paid and had he known this it could have been done long before. Blackwell and William were delighted to hear that their weapons would soon arrive but less delighted when they read on and found that the duty was about £30, well over 100 per cent of their value. A further charge was levied some weeks later when William went back to collect his rifle and ammunition that had previously gone missing.

Tuesday 11 March

Blackwell returned without the guns and said they were unlikely to appear for some time. The party moved on to the eleventh camp, at kilometre 70 on the railway line. The men reported having seen lion in and around the camp the previous night and they were extremely nervous because they had no means of protection. Usually, they spent the evenings singing and dancing to the accompaniment of a *batook* (a drum made from an old barrel with a hide stretched across one end and a stick hanging down the centre). Now they were too frightened and kept very quiet, maintaining bigger fires around their huts to stop the lion coming in.

If that were not enough in the way of problems another serious one arose because the man in charge of the railway siding refused to let the campers have any water. They had already received permission to have as much water as they required from one of the directors of the railway and after some discussion the matter was settled satisfactorily.

Thursday 13 March

William annoyed most people in the camp by setting his alarm to go off early and he had everyone awake and busy long before daylight. The food boxes were completely empty and the fresh supply sent by train from Lobito failed to arrive. Luckily, some of the men had a small supply stored away, which lasted the day.

For prospecting it was another blank day, but the lack of

minerals was made up for by finding fresh spoor of elephant, leopard and buffalo, and they saw seven kudu in quite close proximity. MacKay saw the same tracks around his camp, but he was fortunate to have a rifle supplied by the company.

The whole area they had passed over, especially at the foot of Mount Sohoa, was reported to carry gold and the tracks of other prospectors could clearly be seen. Despite much panning of samples, mainly in dry creek beds carrying likely looking sands, and also many chippings of blue quartz, there was not a sign of gold. It was concluded that the reported gold was probably lepidolite, a quite frequently found gold looking mica.

Along the limestone belt where the camping party worked the mornings were getting hotter and by 4.00 p.m. quite a strong breeze had sprung up from the west, which was pleasantly cool. The nights were cold enough to need a blanket in addition to the sheet fleabags. Apart from the common fly, insects were not too troublesome and there were hardly any snakes, but there were many small scorpions.

The proposed programme of work was to have an early breakfast, go out prospecting as soon as possible afterwards and return to camp for lunch between 2.00 and 5.00 p.m. At the same time a compass survey was undertaken, a task that was simplified greatly by having the railway survey map with them. Work in the afternoons was more difficult to plan because of the wind, which would have made the use of a blowpipe (which Mills had forgotten) – for identifying minerals – very difficult. This work was undertaken with the use of William's balance and specific gravity tests, which proved satisfactory when used. They found no trace of the copper they had hoped to find or the reported gold.

The light thorn bush country took a bit of getting used to and at first William's clothing got badly torn. The bush was rarely thick enough to need cutting but it was not always easy to find a clear (and safe) way through; returning to camp was often by a very indirect route.

Saturday 15 March
By now they were all very hungry; no food could be purchased locally and all wires sent down the line produced nothing. They managed to borrow a day's supply from MacKay but this was not enough and it was not possible to move forward when the men were all starving.

Sunday 16 March
Early in the morning the train arrived at last with the food and the gang was relieved to be able to have a good meal. William and Blackwell thought they should discuss future moves with MacKay and shortly before 4.00 p.m. they embarked on a strenuous walk of two-and-a-half hours to MacKay's camp arriving shortly after dusk.

Being so late MacKay had given up thoughts of them arriving that night and anyway he was short of water. They only carried their beds and essentials for one night, so Blackwell and William shared a grass hut as a bedroom, which was probably a good shelter from the sun by day but proved totally inadequate protection against the heavy rain that night. They sat down to supper at 10.00 p.m. and went to bed well after midnight, finding it most disturbing to be woken by rain at 3.00 a.m. It was a restless night. William tried to sleep under his sun umbrella and a mackintosh but could not, so he crept along to MacKay's tent where he slept on the ground.

Monday 17 March
In the morning they set off to return to their camp again arriving back shortly after dark, just before another exceptionally heavy storm broke. Although both their tents were pitched on sand they were flooded out during the storm.

Tuesday 18 March
After drying the tents in the sun they got off to a late start

at 9.30 a.m. They walked about eight miles and had now left the sedimentary rocks and were in schist country. (This is crystalline rock with its components arranged in layers.)

The workers had begun to cheer up a bit at the last camp but during the day they became more nervous than ever and the night passed in absolute silence, the wind and rain adding to their depression. Their tension was further exacerbated when out walking in the evening they came across a leopard that showed strong resentment at their presence and caused Blackwell's two dogs to come hastily to heel. The railway line was heavily marked with fresh kudu spoor and there was an abundance of small game.

They pitched their thirteenth camp close to Coruteva station. The village consisted of just a few local huts that met the labour requirements of the station. It was difficult to understand the need for a station at this point except for a disused farm two miles away.

Wednesday 19 March

The world had become a much brighter place in which to live. William's case of whisky from MacDonalds had just arrived, having taken three weeks to do the last 51½ miles, including the last 18½ miles that William's carriers did on the last day.

After yesterday's excitement with the leopard it was decided that William should get the train to Lobito next morning (there were only two a week) to fetch a rifle, either William's, Blackwell's or a hired one, to protect them.

Thursday 20 March

William arrived as early as 9.45 a.m. but Colonel Greenwood was away and nothing had happened since he had advanced the customs duty. After much hard work he obtained a promise that Blackwell's guns would be on the train to meet him at Benguela next Saturday, 22 March, and his would follow soon after. William had the foresight to hire an old rifle and ancient ammunition at an exorbitant price from Manham, which was just as well as Blackwell's guns failed to arrive at Benguela as promised.

Sunday 23 March
The acquisition of a rifle did a lot of good and after the men heard the firing of a single round they cheered up and resumed singing in the evenings. William went out in the morning with his rifle to get some fresh meat for the carriers to eat. Near the camp he came across two leopards, one on either side of the track he was following. At close range, but out of sight, William decided not to engage in an encounter with the leopards as the ammunition supplied was unlikely to go even moderately straight except at the shortest range. The leopards and William got off unscathed.

There was another food shortage because the boxes that had been on the way for some days had still not arrived from Lobito. Sometimes cars would get up the same day and others took weeks to do the journey. They were able to get some fresh milk from a small herd of cattle in a compound near the station, but they got severely bitten by common flies. Moving on had to be delayed again and this time their time was taken up fighting white ants by painting the food boxes with a creosote preparation that subsequently added its own special flavouring to each meal. While sitting in the grass huts in the evening it was possible to hear the rustling of the ants as they moved away in the dry grass.

Wednesday 26 March
MacKay joined William and party again and told them he saw much fresh lion spoor on the way up. It appeared that the pride of lions already encountered was moving back to the sedimentary area William had recently covered because the rains had produced fresh grass there and no further traces were seen in the following days.

Friday 28 March
Friday brought them to camp 14, which was near a road to

the north of the railway and halfway to the Cavaco River. Here the bush was very thick, especially near the river, and might have been good country in which to find kudu but very few were seen.

Sunday 30 March

William was determined to do some shooting now that he had a rifle and he set out towards the river and a deep wide valley to the northeast. After only going a very short distance his mind became rather preoccupied with the red-hot trail of a lion moving in a northerly direction. William's gun carrier considered the spoor was less than an hour old and suggested the best direction to look for game was to the south. William had other ideas and, thinking the spoor was about four hours old, felt there was a good chance of finding the lion when it rested in the heat of the day. The two men followed the spoor for over two hours but the lion had now got back to his mate and one nearly full-grown cub.

As the rifle had not yet been tested William was not going to take any risks, so he took a wide detour in an easterly direction to return to camp. After two hours they had descended into a valley where a herd of female kudu was feeding and almost sleeping. One was broadside on at about 60 yards. William wasted no time getting down on one knee for his first shot, but nothing more than a feeble click happened. He hastily reloaded, by which time most of the herd had seen him and started moving. One 'special' one turned round and for about a minute gave a full view of her other flank. The target seemed as big as a haystack – but again a misfire. Before he had reloaded for a third time they had all gone. William was furious with Manham for supplying such dreadful equipment and at such a price. He followed in the direction of the herd and suddenly came across a large bull kudu feeding. For a moment they were both too amazed to move and just goggled at each other. The kudu pulled himself together, snorted and bolted off while William, at a range of about 20 yards from above, selected a fatal spot and shot. The recoil was barely as much as from a .22 rifle and if the animal had been hit the bullet would probably have bounced off. William followed the trail for a few moments in the hope of finding blood, but there was not a drop to be seen. He returned to camp feeling more than a little pleased that he did not get on shooting terms with the family of lion.

Blackwell spent the day returning to Coruteva station in the hope that the other guns had arrived, but they were not there. Instead, he was told that martial law reigned throughout Angola and there was no chance of obtaining the weapons until that ceased.

Monday 31 March

William and Blackwell started the day by covering much of the ground he had gone over the previous day but returned by way of a wide detour. The first five miles were easy but the remaining twenty before returning to camp were hard going, though luckily it was a moderately cool day. Much of the ground was loose sand covered with tufts of grass three to four feet high. Again, William had the chance to shoot a fine kudu bull at 150 yards. He took the shot sitting down with nothing to impair his view, yet again he missed. It just had to be dreadful ammunition. William and Blackwell followed the animal's trail and found him again only 30 or 40 yards away. Blackwell was too surprised to take a shot at only 15 yards. For the rest of the day they saw numerous bush fowl but did not take them on with the rifle.

On returning to camp William's back became very sore and whatever the problem was he was confined to the camp for a few days.

Tuesday 1 April

Blackwell continued prospecting for minerals and looking for meat. He put up a panther in thick bush at about 20 yards, and it followed him and his dogs all the way back to camp. He was terrified and the whole camp was very disturbed all night, but at no time did the panther permit the chance of a shot to relieve their anxiety.

Wednesday 2 April
While William was out for a bit Mills turned up and told Blackwell that martial law had ended and that the guns should arrive within a few days. He then went on the hour's walk to MacKay's camp where he spent the next few days, much to the relief of William and Blackwell. On Thursday, Friday and Saturday William was only out for a short time because his foot had become septic; he now had a high temperature as well as the back pain. On the Thursday and Friday Mills took Blackwell out in the lorry in the afternoon on a shooting expedition with borrowed weapons, producing nothing in the way of food. Blackwell did manage to produce a klipspringer antelope, which he cut up. He sent one half to the other camp, which was also out of fresh meat. This generosity proved very fortunate because during the night Whiskey, one of the dogs on loan to Blackwell, helped himself to the entire supply despite being very sick in the process. Blackwell then managed to recover enough from the other camp to keep them going for the time being.

Friday 4 April
William was very upset when Mills again turned down his request for expenses for a trip to Lobito, especially after having allowed those incurred by Blackwell and Frazier. He then wanted William and Blackwell to accompany him to Lobito to get cash and guns, but when the choice appeared to be William Mills objected, knowing he would be in for a hard time. William therefore sent him a very rude letter about the expenses, with an accompanying one to a higher authority. These were to be delivered to him before he got out of bed the next morning (Saturday). This resulted in Mills being upset for the whole day. Blackwell returned that night with a very polite verbal message from Mills requesting that William withdraw his letters. Blackwell brought his

guns and promised William would have his the following weekend. He also brought back verbal messages and an astonishing letter from Colonel Greenwood indicating that steps were being taken to remove Mills from his job. On hearing this William withdrew his request for an enquiry into why his claim was not treated in the same way as the others.

Monday 7 April
William was offered a shotgun and went out for a strenuous 15-mile walk into the sandy plains. He returned to camp by 2.30 p.m. hot, exhausted and ready for many cups of maté.

Tuesday 8 April
It was a very disturbed night with leopards hanging around because there was more fresh meat hanging up. Blackwell and William concluded that it was no longer safe to stay there and moved to camp 15 on the Cavaco at the foot of Mount Assango. They were short of carriers because a number were being loaned to MacKay. The situation remained a little tense and chaotic and during the day Blackwell and MacKay exchanged notes a few times, on each occasion changing the previously made plans.

MacKay predicted that the rains would now stop and soon there would be a water shortage, and that unless a hasty start were made the projected crossing and prospecting of the ground to the north of the river and west of the limestone–schist contact would be impossible. He refused to believe that water would be found for months to come in the river sands, even if no more rain fell. He proposed that they should all join up at the foot of Mount Assango as soon as possible, move downriver to a point clear of the schist in two jumps, and then out into the barren areas as quickly as possible. All three gangs would be required to ensure an adequate water supply.

That night Liza and the other camp dogs howled away warning that leopards were around. Late into the night William got out of bed with a pistol and sent a shot off making sure that he missed the animal by a respectful margin.

Thursday 10 April
MacKay arrived back at William's camp ready to start the walk down the river. He was surprised to find pools of water seven or more feet deep and, according to scouts sent ahead to find the route, it was impossible to get down the river. MacKay would not admit he was wrong and, after trying to be on Mills's side in discussions, was surprised to hear that Mills was likely to be recalled soon and realized he had backed the wrong horse.

A ton-and-a-half of food at Coruteva station was under guard, and a little more prospecting was undertaken that resulted in unexpected findings of sulphide copper ore, or chalcopyrite, in a number of boulders near the camp. Shooting was also more successful, seeing a zebra and bringing in a kudu. There were large numbers of baboons round the camp; they never molested anyone, but they added to the disturbance the leopards caused at nights.

Saturday 12 April
After three days of rain the Cavaco had risen considerably. Being unable to travel down the river bed they started the march via the railway line to its point of debauching from the schist. They reached Coruteva station camp that night and in the evening William went to the station to get some bread. The camp itself had fallen into a very bad state because the water carriers left behind destroyed many of the huts for firewood and white ants ruined the rest. They also attacked the tent pegs making life very difficult.

Sunday 13 April
William's foot was bad so he rested up before the next walk. It was another wet day and three gangs moved six miles down the line before turning northwards. Messages were sent to Mills advising what they were doing, asking for food urgently, and for William's guns, offering to pay all

expenses for a lorry to bring them up. William waited at the road–railway crossing near Coruteva for the goods to arrive. Early in the afternoon Mills sent a wire saying that he wanted a gang for himself, or one of the three with a gang to remain at Mount Assango for him to come and see the copper find for himself. Blackwell and MacKay refused to stay, knowing that Mills was not keen on being there alone with William.

Monday 14 April
It was a chaotic morning, all the carriers having been sent back for final loads, and when they returned the gangs had to be separated and stores divided as far as possible. MacKay and Blackwell went off with their carriers to continue their route and William returned to the station for a third time. Things were getting desperate because for two days the camp was out of bread, vegetables, whisky and other things. Still, the urgent stores failed to appear and, since he was in comparative luxury, he gave nearly all his essential stores to Blackwell who was on the move.

Thursday 17 April
In the evening the stores arrived in a huge packing case and in the food packages the sugar and rice had scattered everywhere. The sugar at best was a dirty white colour but at a price of one penny a pound it was a bargain. It was now too late to divide the supplies before going back to Mount Assango.

William's guns also arrived but they were in a filthy condition and very rusty; his shotgun ammunition had been changed and he received some unsealed rounds all of one size. None of his buckshot and lethal balls had been sent. It turned out later that Mills sent Blackwell's ammunition to William but when the error was discovered William's ammunition had still not been sent on and it was directed to him again. When it did arrive he found he had lost most of his heaviest shot and half the lethal balls.

Friday 18 April
It was another wet day and William moved back to Mount Assango again.

Saturday 19 April
At last William had time to sort things out a bit and get comfortable. Some men were sent to Catengue for bread and vegetables and others for milk and eggs. Blackwell was sent his requirements so long as he could withstand the wet weather or being immersed in water because the route there involved crossing the river frequently and could only be managed if the carriers had light loads. He also sent them some fresh meat, which they needed. This route enabled the men to keep in touch for a few days despite the Cavaco having risen still further.

Sunday 20 April
William was up early and, after sorting his things, set out at 9.00 a.m. to Mount Assango. He made a thorough search on a flat at the foot of the mountain separated from the camp by a dried-up river bed. Quite early on he had a shot at a bushbuck that got away, but later took a long shot at a steenbok. He got back to camp at 5.00 p.m. for lunch. Thinking that this was a bit late for lunch he decided that in future he would take lunch with him, including a table and chair, and even contemplated taking his gramophone to play some music during a short rest after lunch. However, all these lovely thoughts were but a dream.

Monday 21 April
It was Easter Bank Holiday Monday and Mills's arrival disturbed William's brief spell alone. He had sent most of his carriers to Coruteva for some more *fuba* (crushed mealies, the local staple diet). Mills met the men on the road and collected all of them to carry his baggage down to the camp. William was by now out on the mountainside enjoying a pleasant day and thankful that all the drinks, except plain water, were locked up. When he returned to camp for lunch

he expected to give and receive a cold welcome from Mills. After tea Mills said he wanted to see the peculiar dyke William had described. It contained dolerite with a breccia cemented with ironstone. Despite his sore feet he took his gun with him and soon he was lured across the river after a bushbuck; he remained out until dusk when he returned with a steenbok after expending about eight rounds. He never went to the dyke and William only waited there a short while before returning to camp for his evening bath.

That evening, while taking drinks and quinine, Mills, quoting himself as an example, told William that everybody in the field should reduce their loads. He said that for a week's trip he only needed 11 pieces, not realizing that it took 25 men to carry them in and 19 to carry them out on his departure. By dinner time the tension was so high that although they shared the same grass tent, they each had their meal at a separate table, Mills being the lucky one to have had his meal cooked by the good new cook who had replaced Mensa. By now Mills's voice was cracked and he staggered to bed feeling there was a conspiracy against him and that he was not appreciated.

Tuesday 22 April

Mills was not prepared to do very much in the morning because of his sore feet and so they looked at a few boulders containing chalcopyrite, returning in plenty of time for lunch. He then predicted that he would find something really big in the copper line and after a siesta (when his feet miraculously recovered) he set off with William carrying guns, saying that the morning was for prospecting and this outing was primarily a shooting expedition.

Mills and William had been out for some time and seen nothing, but just as they were about to turn back to camp they came across a small herd of female kudu grazing about 150 yards away. Mills was the first to see them and immediately started shooting. This alerted William who took a shot at one animal with a hit, but after the second round the cartridge case burst in the chamber and he was unable to close the gun again for some minutes. By the time the gun was ready again the animals were in full flight, but William took one parting shot at long range. They followed the animals across very broken country and soon picked up one kudu under a bush. At ten yards range Mills fired two final rounds to finish it off, scoring one hit and one miss, by which time he was exhausted and had expended his ammunition. He stayed behind while William continued the chase with only one barrel that could be used. Soon he caught up with the herd and had a great chance of another kill. He pulled the trigger for the wrong barrel and the kudu disappeared from sight. The two men returned to camp triumphant with their prize and left the dissecting to the carriers.

That night they recounted the triumphs of the afternoon, each claiming they had shot the animal and each certain that they must have put many bullets into it. In the morning they went to see the hide drying in the sun and found that such was the excellence of their shooting that all the bullets entered by the same hole. The mountain was thereafter referred to as Hill 60 to commemorate the battle fought there, Mills having expended over a dozen bullets and William three. After drinking the health of the poor victim several times, and further discussion about not receiving any thanks for getting the guns from the customs, Mills retired to bed in a very maudlin state.

Wednesday 23 April

There was a slight drizzle in the morning so William extracted the broken cartridge from his rifle before going out to Hill 60 again. They expected to have an 'interesting' morning but although it was very wet they saw nothing. In the afternoon William went to see how some pitting was going on and brought back another steenbok. Meantime, Mills said he had to do some accounts, which he apparently did in his sleep because William found him asleep before he left camp and he was still sleeping on his return.

Thursday 24 April
After a relaxed morning Mills and William each went their separate ways in the afternoon to collect meat to be taken to Lobito. Mills had decided that he was not going to find huge copper fields or anything else here and there was important work for him to do elsewhere. William returned with one steenbok and Mills, having run out of ammunition, returned empty handed.

Friday 25 April
Before leaving Mills gave William several lots of different orders to carry out, which finally boiled down to leaving him alone for a month. He promised to send food and stores but they never came. He left in the morning taking nearly all the carriers with him for the day.

5

The Cavaco Trip

William was now left on his own to carry out the instructions Mills had given him. He was to complete the prospecting down the river to join up with the area covered by Blackwell and MacKay. The track was shown on the railway map, but Mills thought the map was probably done by guesswork and that no white man had ever been there before. The rains had eased off into just occasional showers and the river level had fallen, thus making the task easier. After this he was to move upstream until he reached the railway at Coimbango and this proved to be much more challenging.

26 April to 10 June 1930

Saturday 26 April 1930
William and his men set off early for a ten-hour search downstream on a hot and tiring day. The lower part of the river was closed in by hills of about 2000 feet high, and a slope of 30 to 35 degrees, all heavily covered in thick bush. To utilize the carriers to best advantage he sent some of them up the hills on either side to collect rocks from the summits and others went downstream in search of game, which they required for food. They returned to camp hot and tired with no great finds.

Sunday 27 April
Sunday was to be a rest day but as herds of kudu had been

19. The Cavaco River.

reported William went out looking for them although he did not see any.

Monday 28 April
William moved camp to the best site he could find for a long stretch, which was about 16 miles downstream, leaving most of his personal kit and stores behind. He took a few men with him and the rest he sent on another walk to Coruteva station to collect some more food that should have arrived. By now the food situation was desperate. There was no fresh meat and no game had been seen in the area. The bread that had been kept moderately fresh in a tin had now turned bright green. William declined to eat it, but the men were so hungry they were fighting for it and wolfing it down. There were hardly any potatoes left and the wine was almost finished.

Tuesday 29 April
An urgent messenger was sent to get some food while William continued to search vainly for copper, going 24 miles downstream.

Wednesday 30 April
By evening William was ready to return upstream to an old camp once again. At that moment he received a message from the *capataz* (foreman) that no *fuba* had arrived and he awaited further instructions. As the *capataz* was 18½ miles away it would take a couple of days for those instructions to reach him, during which time he would have plenty of free time, about which William was not too happy.

Thursday 1 May
William had no intention of sitting around waiting where he was, so he took the five men retained and his personal staff to move back to Assango. The carriers started before day-light and managed to have second loads in camp by noon, having avoided the hot part of the day, but still they were exhausted. The personal men were furious at having to carry loads in addition to their normal duties, but little was left to be collected the following day. Immediately on return to Assango William went out with his gun and shot a steenbok for food. There were plenty of them there until they thinned the numbers down over the next few days.

Friday 2 May
During this return move William's only screwdriver got lost. He sent his personal man to look for it but he had no luck. The carriers had got back to camp with the food so now William had all his carriers with him and he sent a few to collect the last items left behind downstream at the previous camp. The paraffin had failed to arrive, which meant that the evenings would be spent in total darkness. On the Friday and Saturday he sent all available men out looking for a possible track to get past Mount Assango. The first three miles were easy but the next section was so rough it was hard to get along even without loads and with full loads it would have proved impossible.

Sunday 4 May
It had been a hard day and in the evening while William

was taking his bath he was disturbed by a snake coming into the tent and going to ground under a box. William called Macai (his personal servant) for assistance, but as soon as he realized what was happening he beat a quick retreat. His assistant then came to the rescue and while William finished his bath in comfort he stood by armed with a big stick in case of need. William realized he would have to root out the snake himself, but was not particularly keen on doing this as he was draped only in a towel. He managed to flush the snake, which turned out to be a small venomous puff adder that subsequently died.

Monday 5 May
William made another unsuccessful attempt to find a route past the mountain but Mattheus, his *capataz*, reported that he, with some carriers, had found a possible pass on the mountain, which, although difficult was preferable to the alternative of going round via Catengue; this would have taken a week as every man would have had to move two loads and there would also have been a water shortage.

Tuesday 6 May
Taking the first loads with him, William set out via this route to select a camp site. The climb of about 2500 feet on a hot day seemed endless. The reports of huge herds of game turned out to be a myth but they did see a pair of steenbok and a few birds.

He returned early to the old camp and went to look for some fresh meat. He hit a steenbok that was standing on a rock on a very steep part of the mountain overlooking the river and, as a result, it fell and rolled at least 40 yards. The carrier went ahead and as William followed he dislodged a huge boulder, which rolled down after the man, gradually gathering speed. After William's warning shout the man barely had time to get out of the way and the rock, which rapidly developed into a small landslide, cleared everything away from before it and carried the newly slain meat down a further 200 yards, almost into the river. It was a difficult and exciting descent before reaching it again and every care was taken not to start another cascade of stones.

It was getting late and it seemed the easiest route was to go right down into the river bed to get back to camp. After reaching the river the route proved more difficult than anticipated and soon William tried to climb out from a gully. It was a particularly dangerous climb and at a most inconvenient moment William heard a sharp hiss from behind his shoulder. He was in a precarious position and dared not look round for fear of a big fall and he could not spare a hand or foot with which to defend himself should the snake try to strike. All he could do was continue to climb up the rock face hoping that if the snake did strike it would only hit his hat and do no more damage unless it fell and broke its neck, which it did. Soon William was back on ground where he could walk easily again.

Dinner was late and after William had been back in camp for a couple of hours his cook Lucian came to ask him for some tinned meat for dinner. After all the difficulties he had had getting fresh meat William gave the man a piece of his mind, which he felt was fully justified. By nightfall the weather had turned cold and raw.

Wednesday 7 May

It was still cold in the morning and the carriers were reluctant to rise; consequently, breakfast was late and much of the climb was done with the sun well up. The going was tough and it took the carriers four hours to cover two-and-a-half miles. Macai got left behind and so failed to see where the carriers turned off to climb from the creek bed. He got completely lost and turned up several hours after the rest were in camp.

Camp 22 was very dull and the main interest was to move forward into more interesting country. There were few animals to be seen but William did get a couple of hits on kudu and one on a zebra after a long stalk. The zebra galloped away as though untouched, but half an hour later it was seen tottering past the camp. William had heard that zebra skins make the best hide obtainable and he really wanted this one. William was concerned that his shooting at these large animals lacked confidence as a result of the hired rifle and many suffered from receiving the bullet too far behind the shoulder to be effective.

William found more traces of sulphide copper in this region but insufficient to warrant further exploration. The sedimentary rock was interesting in that it occurred so far in the igneous (volcanic) country. The country beside the river was precipitous and often covered in thick scrub. After an hour's hard cutting they only advanced a few hundred yards and suffered casualties when two men were blinded for two days from the juice of a cactus that was almost as poisonous as the ones in Nigeria. After this William thought the river bed might be safer, but sometimes the bottom was firm and they could move rapidly and then without warning they were liable to walk into loose sand, dropping as much as shoulder deep as if they had stepped into water. In other places boulders were thickly strewn about and provided hard climbing. After a mile or two it became obvious that it was a hopeless route for carriers with loads.

In the evening they came across a man grazing cattle and he said that there was no route forward other than by going back to Catengue.

Saturday 10 May

William was not going to retrace his footsteps and, being determined to find a way across Mount Assango, this time searched out a route about a mile from the river. He made his camp a little over an hour's walk from water and seemingly in the right direction. The camp was delightfully situated on top of a spur with a steep drop of about 500 feet below and commanding fine views of the surrounding countryside. A man kept a continuous lookout for game but did not see any. Being very low on paraffin William lit a big fire, which in the full moon gave ample light for his bath and dinner. The one drawback to taking meals by moonlight was that salt and pepper were hard to see; an excess of red pepper with bush fowl was painful and on one particular occasion it was necessary to drink much water to cool the mouth down. After a hard day, bath and dinner William relaxed by playing some records on his gramophone and was lulled into pleasant feelings of peace and comfort.

Sunday 11 May
A walk after game included climbing another high hill and this provided hopes of finding a way forward.

Monday 12 May
William set off with some carriers to prepare a new camp. They climbed two large hills and had a long scramble along the side of a third, from which they could see the valley of the river opening out a little further on. Mattheus went on ahead to prepare the camp while William mapped the country. Soon there was a great commotion below. William rushed down and found that a nine-foot water python had been killed, its long coils continued wriggling and none of the men would go near it. Once sure that it was dead William jerked it hard by the tail, flipping it back among the men. They scattered with screams, soon turning to laughter as they realized they were out of danger. The snake was later skinned but the skin disappeared. They all returned by different routes in an attempt to find the easiest way to carry the loads next day. This time William was the last to get back to camp, getting in at 5.30 p.m., but felt satisfied by having fitted in a lot of genuine prospecting on the way. Also the scenery was wonderful, although the going was often very difficult for long distances.

Food was again running out and urgent messages were sent to Lobito to get more out quickly. Another man was sent to try and get some paraffin for a night light.

Tuesday 13 May

William moved forward to the new camp taking about three hours to walk five miles. The last part of the day was impressive, being down a steep drop of about 1000 feet with a slope of some 40 degrees. Often William needed his hands for the descent, and it seemed a marvel that the men got down all in one piece and not a load was dropped.

On arrival at the camp William was not pleased to find his tent was out in the open because in this hot weather he wanted it under the biggest and thickest tree available.

Wednesday 14 May

William went out to get some fresh meat and took his shotgun with him. While walking back in the river that evening he heard the sound of wild pigs. The thought of a change of diet was appealing. One was hit with a lethal ball and struggled back into thick reeds. There was a squeal or two and then all was quiet. William and the men went in search of the pig but never found it. Afterwards William heard that all pigs in Angola, wild or otherwise, were badly diseased.

Thursday 15 May

William spent a normal day prospecting in the morning and early afternoon and then packed for the move after lunch. He tried a new programme for the move and had the camp prepared in advance. At 4.00 p.m. he went forward to the new camp. The Cavaco valley opened out to become nearly a mile wide and walking was easy, but the sides remained steep and covered in thick bush. Schist became rare, gneiss (coarse-grained metamorphic rock of quartz, felspar and mica) being more frequent, but all the surrounding hills and country to the east were granite with rare diorite, the big rounded domes being very typical granite. Much of the valley was used periodically for cattle grazing and milk and eggs were obtained here.

According to the carriers there was plenty of game in the area so William walked in front of the column of carriers.

20. Baobab tree, 59-inch circumference measured eight feet from the ground.

While walking through cactus country William again had his shirt and shorts ripped to pieces beyond repair. Shoes were more comfortable than boots but this meant that socks were quickly ruined by short clinging grass that soon covered any woollen article it touched. Walking for five minutes was enough to form a thick mat on the socks or stockings and this gave the men half an hour's work to pick them clean again. Despite forecasts, there was no sign of game but a little fresh lion spoor gave reason to hope for some shooting. By dusk he had his tent up and everything stored away.

Friday 16 May
Stores were brought up and the requisite pitting done, which allowed for an immediate move forward. This was also essential because of the food shortage and being so far from the nearest railway station.

Saturday 17 May
Mattheus went ahead to select the next camp site, but when

21. In the granite plateau.

William found that this was only half an hour's walk away he insisted on walking a further five-and-a-half miles. During this walk he shot a brace of small buck with a shotgun and this provided food for the whole gang for one day. Having eaten well William treated himself to some music after dinner.

Sunday 18 May
William had heard there was some game in the area and one of the local cattle men offered to act as a guide. They left after breakfast at 8.00 a.m. (much later than William liked), but saw nothing and William returned to camp at 12.30 having become blinded in one eye for the last two hours by a twig. In the afternoon mail arrived with news that food for the gang was on its way by train and would arrive soon at the next station ahead.

Monday 19 May
This was a bad day. William started out by prospecting the northern side of the valley and the granite bluffs beyond. As soon as they reached the slopes cutting had to begin and for

a long time progress was slow. Cactus milk continuously dripped on all the men, there was thick thorn to negotiate and, to make matters worse, a number of snakes. Eventually, they came out on top of a big plateau covered with light scrub. Keeping to the bare granite for a mile or so William selected a nice shady spot for lunch and while that was being prepared he went into the bush. There he saw some game, which he thought might be duiker, and a few birds. The pitted creek beds showed nothing of interest.

After lunch William mapped out the country and sent all the men back to camp. He then went out to circle a small hill and returned at leisure. There seemed to be plenty of time and game, and cattle tracks were plentiful. Luckily, William did not dawdle, for after an hour the tracks petered out and he found himself in thick bush and long grass. It was almost too late to turn back and the ground appeared to open out again a little further on, so he set fire to the grasses behind and went forward, mostly doubled up under the shrubs and following an old animal trail. Occasionally he came out into open grass that was very stiff and dry and it was difficult to decide whether to push through it or try and scramble along on top. After being doubled up he now had to step high, often making two, three or even more efforts for each pace forward to get a foot on top of the grass, or at times putting full weight against it to shove it aside. After each bad patch he set fire to the grass and with a favourable wind the flames went roaring up the valley making a loud noise. The going was extremely hard and by sunset he had barely started on the descent into the main valley, it being impossible to take a short cut because the rocks on either side were too steep to climb. There were also many baboons in the hills and William had doubts about spending the night with them.

Just before dark his luck changed and he came to a long spit of rock down which he could run and scramble fast and now he felt fairly confident of getting back to camp safely that night. He had to leave the easy going of the rocky spur and was back again in the long grasses. Each time the place seemed suitable he started fires that went off with a great roar and crackle and made him move a bit faster to get out of the way as the wind had changed and the flames were now chasing him. As it got darker the route opened out a lot and the scrub gave way to more thorn; soon William was dripping blood from his many encounters with various types of cacti, especially the single spiky cactus grass, which severely damaged his legs and of which he had no warning until he actually walked into it. Sometimes he managed to get onto a track for a few yards, but it was too dark and it was soon lost and he was back into another belt of prickles. There was no moon and much of the time it was hard to make out even the bare outlines of the hills ahead, but somehow William got enough outline to keep his direction.

At 7.30 p.m. William thought he heard a rescue party calling and in reply to his shouts he heard much rustling around him. He let off a single round from his gun and that started a fearful commotion because he found he was in the midst of a large number of baboons. He quickened his pace considerably and let off a second round, which was sufficient to make them keep their distance. Twenty minutes later, when nearly back in camp, he met the search party coming out with lights. All he wanted now was a drink, bath and dinner. Lucian, the worst cook William had ever had, chose this of all night's to produce a particularly badly cooked meal, for which he had his pay reduced.

Tuesday 20 May

By morning William was stiff from struggling through the long grass and the fires could still be seen blazing well. His clothing had not suffered too badly but his shins were raw.

The carriers went on ahead and Mattheus, having been told off for not going far enough on one occasion this time went too far, for he chose a camp site about three hours' march ahead along an easy cattle track, but William did not relish going more than two hours. Mattheus was also in charge of rations and they were running out a week early, which was another reason to move on as far as possible. William stayed back for an hour to finish work on the map.

Wednesday 21 May
William was called at dawn because guinea fowl had been seen in the river bed. He quickly dressed and after an hour brought in two birds after five shots, which was below his normal standard. Most of the men then went with Mattheus to collect food from Coimbango and the few remaining with William would have to go without food until the others returned three days later. They found a few crops about on which to survive but less work was required of them under these circumstances.

Thursday 22 May
William walked about 12½ miles a day alone. The walking was much easier, the valley was more open and it was no longer necessary to have a party of workers cut a path ahead. The one exception was at the bed in the river where he attempted to go northeast to the Cubal River but was beaten by thick bush.

Friday 23 May
Another 12-mile walk and this time he came upon a hill requiring much cutting and when clear of this he reached a granite slope where progress was slow and difficult.

Sunday 25 May
A messenger turned up from Blackwell who said that Blackwell had been looking for him for the past 11 days and it was only when he came across the tracks of the men sent for food to Coimbango that he was able to find him.

Lucian, the cook, was getting slower and his meals were worse than ever. On arrival at camp 28 he took an hour to open a tin of meat, though he had a perfectly good tin opener and appeared to be working all the time. He was not built for speed or for life in the bush and William decided it was time he went. In addition, his wife was noisy and he had a pack of awful little dogs that yapped all night and most of the day. Because of this noise Lucian had to have his hut far from William's tent, but that created problems in getting him to start cooking meals on time.

Tuesday 27 May
William had seen no kudu for days but today he saw a few feeding. He sent a man back to get his gun but the animals had gone before he returned. William promised the men that he would go out early next morning and get one.

That night the men decided to celebrate the shoot before the event and gathering together with a few locals (they were now in a crowded area with a population of nearly half a dozen men and women); they had a great dance to the *batook*, keeping up the noise until nearly dawn. As a result every animal for miles around cleared out of the way.

Wednesday 28 May
Although William was out well before the light was good enough to shoot there was no animal in sight. The men went ahead and prepared the next camp. William left in the late afternoon and, since it was very cold at night, he had to get the whole gang to collect enough wood to keep the fires burning all night. In addition to the fires the men had to double up to keep warm enough through the cold night. Lucian's dogs were particularly noisy that night, which made things even worse.

Thursday 29 May
Lucian's ideas of breakfast did not please William either. Last night Lucian had said there were eggs and this morning he stated there were none. When told to go and get some he retired to his hut at the far end of camp and, after a suitable lapse of time, returned saying he could not find any. William could stand it no longer and sacked him on the spot. Lucian was pained and surprised to be paid off and

told to move out at once; Mattheus was told to see that he was gone in under ten minutes. The job of factotum fell to Macai and for the next few days William was confident his meals would be punctual and well cooked. In his new capacity he said he would look for another cook and he spent the next few days going to the city in his best regalia and returning somewhat inebriated. The walk back from town was long enough to have him in a state fit enough to produce a passable meal and nothing was said about the episode. There was a certain advantage in his spending his money quickly because there was less chance of him going out again when it would be less easy to replace him.

William did not like this camp being sited so close to the road into Coimbango and in the morning he went back into the bush country. Guinea fowl were about but by now the ammunition that had not been boxed had deteriorated so much that it could not be used.

As usual, there was a muddle in the orders William had received and Mills, having forgotten about the last part of the trip, was expecting William and his full gang back in Lobito three weeks earlier. Blackwell was waiting in some unknown camp for William to appear and they were to join up once again to work up north along the sedimentary schist contact.

William got his things together and went down to Coimbango station. His gang was to walk with light loads down to Catumbella to meet him there on his arrival by train a few days later.

Saturday 31 May

Camp 30 was a bad move because the site had not been cleared and shovels and bush knives had been packed ready for carrying to the coast. There was only one possible place on which to put up the tent and while William watched the clearing of grass he stood in the midst of a column of soldier ants that quickly saw to it that he did not remain there for long. On examination, William found they were returning from invading some white ants and carried many prisoners with them. William decided to leave the tent where it was and to kill the ants with hot water. Four cans killed thousands, but thousands still remained. He then tried burning off the grass but the ants still continued to walk forward along the column of corpses. As they did not enter the area cleared for William's tent he hoped all would be well and for two days he kept on the lookout for them and maintained a fire at the head of their column to try and prevent them moving forward. After this they suddenly disappeared.

The main body of the gang started their walk down to Catumbella while four remained to help William finish his personal work and the last bit of prospecting in the district. On the first day in this final camp he had a little difficulty getting fresh milk, eggs and fruit, but after paying for it the supplies were sent in all too readily.

The final map was a difficult undertaking because it had to be copied in a high wind. The area covered measured something over 50 miles long and varying from three-and-a-half to eight-and-a-half miles wide. There was a problem in reconciling this with the railway map, which was easily explained when it became known that the surveyor had lost his notes at the end of his trip years before and decided that it was easier to draw in the river's bed from memory than to go back over that country again. He also considered it probable that no one would ever go there again to check it.

The few nights at the last camp were cold, though the days remained hot, but not as hot as they had been during the previous few weeks. The mornings were very misty, spoiling the chance for long survey shots with the compass back to Mount Assango and the bend of the river near camp 27. Hunting in the district was poor because of the number of local people near the station.

On the day of departure William's baggage was sent to the station in the afternoon, but the station master refused to accept it. There followed a period of wild telephoning to get permission for it to be sent by that train and by 7.00 p.m. William was told he could travel with it. The train was due at 4.00 a.m. so he tried to make himself comfortable by sitting in his deck chair in the station master's office. He had a disturbed night with mosquitoes, continual traffic and telephoning and, by 3.00 a.m., he was up and longing for the train to arrive. Having obtained a new cook who was said to be good in the bush his staff had now increased by one. The cook accompanied William as far as Catumbella and deserted after one night. He returned home at his own expense but had the cheek to ask for three days' pay.

William looked out for his carriers on the journey but did not see them. On arrival at Catumbella only Mattheus was there and he said that the others would arrive in about two hours, Mattheus having taken the train as an easy option for the last part of the 'walk'.

As soon as possible William went up to the hills overlooking Catumbella and camped there for a few nights before taking the train back to Benguela. There he tried again unsuccessfully to get the balance of his ammunition and lost gun back from the *materiales de guerre*, and to get some more stores from Benoliel, the Portuguese man who proved to be the most reliable person for sending out extra provisions to the bush.

Friday 6 June

William was back in Lobito. He arrived the day after Mills had returned from a trip with MacKay during which he had had a few minor accidents. In fact he was in a complete mess: he had completely forgotten all his instructions to William about how long his trip was to take, and about the accounting and reporting procedures. He had also mislaid William's ammunition and offered a poor substitute. After considerable trouble William found most of his ammunition but not the precious lethal balls or the buckshot.

Saturday 7 June
William sent on a strong advance party with stores to join Blackwell and await his arrival.

Monday 9 June
William paid an early visit to *materiales de guerre*. The man in charge was offended at the thought that there might be an error of any sort and, but for a tip, he would have been unable to obtain any of his much-needed ammunition. By then there was no train back for hours so he started walking, hoping he would soon get a ride on a lorry. Only one appeared and that broke down after a short distance, so William and his manservant had a long walk. Just before arriving the company lorry turned up but failed to see him and returned to Mills for further instructions. William also sent for the lorry but it did not appear until it was too late to move that night.

Tuesday 10 June
In the morning William got away to join up with Blackwell again, the remainder of the carriers followed carrying two days' supplies of food.

William's musings on Angolan labourers

William's view of the Angolan workers he employed was that, on average, they were 'dirty, lazy, clumsy, ignorant and superstitious. Many were cowardly, often lied and some were rogues.' They had a deep-rooted fear of white men derived from several generations of slavery under Portuguese rule. Without that fear William believed they would rapidly relapse into the simplest forms of barbarism and cannibalism, but their fear also created a more attractive side of their nature, which made them likeable people until something went wrong.

[Incidentally, as an aside, it is important for us to remember that the opinions William formed were based on his experiences in 1930 (more than 75 years ago) before air travel made communications between continents easier and when conditions there were very different.]

A number of different tribes were scattered throughout the area covered, the individual tribal areas varying in size from a few square miles to vast stretches of country. Half William's workers were from towns like Lobito, Catumbella and Benguela, while the other half came from country areas. The men rarely exceeded five foot nine inches in height but were taller than men from other districts; according to William, who was proud of his own physical fitness, they were of poor physique and had low levels of intelligence. The men from north of Lobito were particularly weak and few could manage a day's work with a 60-pound load. William had commented that they had great difficulty working with a pick and shovel, that their bush craft was poor with little aptitude for spooring, and that they were clumsy when skinning carcasses. They were terrified of snakes and very few would attempt to supplement their meat ration by trapping birds or small animals, and those who did used very primitive methods. None apart from the hunters who were retained to supply meat to the farms went into the bush alone. This resulted in a poor sense of direction and many lost the camp when it was in a rather inconspicuous place.

Women formed a very small part of the population in country districts and William thought that this seemed to suggest that less value may have been placed on preserving the health of their female as opposed to their male children. Women were hardly ever employed to perform household duties for European men or women. William only allowed three women, including the wife of the *capataz*, to stay in the camp in districts where there was a shortage of water, but when they were near small towns or big farms the number could increase to as many as 20 until either money, water or food necessitated a reduction again. It was unusual for a woman to stay with a man for any length of time, although some of the town men maintained a family at home with whom they would deposit their few belongings and to whom they would send money whenever possible and according to their means. William noticed that young boys joined their father's at an early age and the girls stayed with their mothers until they were old enough to look after themselves.

Like many people from the west coast of Africa they were highly sociable and always ready for their own forms of local music and dancing, which would go on from dusk until dawn whenever permitted. The music for dancing was mainly played on a *batook*, which also accompanied their songs. Generally, the drum was tapped by hand and the sound of a *batook* carried over vast distances. On a still night it could be heard by another camp more than ten miles away and it seemed as if it carried further across limestone than granite hills. If allowed to go on too long the men were obviously of little use as carriers the next day. The local tribes did not appear to practise any obvious form of religion, but one of William's men had been brought up in a mission and he encouraged his fellow workers to sing hymns, though with limited success.

Pay varied according to the district. Lobito men got as much as ten shillings a month plus food, but the men on farms further north were lucky to get five shillings a month. Their Portuguese employers were purported to treat them very badly, beating them on the slightest pretext and, when labour was scarce, withholding payment after a month's work to prevent them moving elsewhere. William paid his men on time and never suffered desertions or loss of loads. Furthermore, he was rewarded by men coming from far away to join him and he found it safe to make small advances to those who had worked for him for any length of time. More than half his original labour force remained with him throughout the tour.

Only the townsmen had any real idea of the value of

money, but they would all quickly learn what their correct pay should be. In the few country districts where it was possible to purchase produce like eggs and fruit, people were vague about what to charge, though their prices were never cheap. The local traders were notorious for extracting money from the men for worthless trinkets. After some months a few of the men would take their goods to William to ask him if they had paid too much for them and by this means they managed to keep their money for a little longer.

Under Portuguese law the workers were supposed to receive a kilo of *fuba* and 200 grams of beans a day, with salt and palm oil for cooking issued each week. Once a week a kilo of meat or dried fish (with a dreadful smell) was issued instead of beans. William's gangs were lucky to receive far more meat issues than other gangs. The issues were large in size and the men had healthy appetites.

For the last five months William never had to buy meat or fish with company money. He also offered to supply all the other gangs with their requirements so long as they collected it themselves. By supplying meat and paying the wages early or on time William was able to have all the carriers he needed. In the last month it was difficult to keep the numbers down because men would walk for a week from Lobito or Benguela in the hope of finding employment with William's gang. The other gangs depended largely on semi-coerced labour obtained from local labour recruiters and such men, apart from often being in poor physical condition, were unwilling to leave their local districts and therefore frequently became immobile for weeks on end.

William was handicapped by having to cover the roughest country in dry and desolate areas with very few women in tow, about which his men, especially those from towns, were unhappy. The other gangs were subjected to fewer restrictions because water was more readily available on their routes.

A man going out to join William's gang always had to

accompany his own messengers and would usually carry their loads in return for his safe conduct to camp. Men moving across country alone with no written authority from a white man were liable to be arrested as vagrants, from which it was a short step to being handed over to work on a farm without pay or being sent to prison. Everybody had to carry identification papers issued by the government to show that they had paid their head tax. The amount paid for this tax varied according to the area in which a man lived. Those living near the border with Northern Rhodesia (now Zambia) or the Belgian Congo (formerly Zaire but now the Democratic Republic of the Congo) paid less as an incentive not to leave the country. Those who were prepared to work for the government on public roads for two months a year without pay, with little food and only what water there was in the locality, were exempted from paying the head tax.

The regulation load was 60 pounds, but this was too heavy for some men. However, some men, especially those who were working for themselves, were known to carry over 100 pounds for a long march each day when the crops being cut were getting a good price in Lobito. On such occasions all the women and children would be involved in the harvest.

All workers were supposed to be decently dressed with loin cloths only permitted when they were deep in the bush. Their suits were mostly home made from old sacks, though a few dressed quite well. When townsmen went into the bush they stored their best clothes at home and after the first day often seemed to have very little to wear.

When their sacking suits wore out they started to buy clothes and the size of their wardrobes increased. With the various other treasures purchased along the way, many soon acquired quite a few personal belongings. At first even an empty sardine tin was considered to be worth preserving and empty bottles were particularly valued because they could be used for carrying extra water over and above the ration for the dry areas covered. English books and papers like the *Morning Post* would be carried around for days by men who could not possibly read them or understand the pictures.

Unlike people from the north, who carried loads on their heads, these people would use two sticks, each about seven feet long, tied together in a nutcracker fashion; the package would be fastened in the open end and the other end would be used as a fulcrum. It was then easy to lift the load onto any man's shoulders and it was also easy to rest the load at any convenient halt or even when climbing. The problem with this method was that many of the boxes were carried upside down and when stored away in William's tent at the end of a march the locks were invariably inaccessible.

William found the local men naturally quite indolent, but they thrived on hard work and fair treatment. Most of the local white men in the bush seemed to be strong believers in beating and frequent fining. William took the line that an occasional heavy fine was justified for a known offence accompanied by the knowledge that a beating might be forthcoming for sloth. The threat was a sufficient deterrent and served to make the men handle his kit and equipment so carefully that not even a glass got broken through rough carrying. Badly packed table and kitchen boxes, however, were apt to make the food taste of paraffin and an occasional tumbler to get smashed.

Personal servants were reliable provided each was made responsible for his own domain. Macai, for example, was the only person allowed to enter William's tent in his absence, and the cook had to answer for every scrap of food issued to him. The personal servants appeared to take as much care as possible to see that if pilfering was possible they alone should be the ones to do it, and for William it was better to be robbed by the few than by many.

Because the cooks' training had been in Portuguese houses

where a lot of oil is used, the food was not always to William's taste; also constant supervision was essential to obtain acceptable standards of cleanliness.

The men had very limited supplies of traditional remedies and often went to William for help with their ailments. Diagnosis was difficult and in many cases the law, which was designed to protect them, made it necessary to discharge men who otherwise could probably have been treated and cured. If seriously ill they were happy to return to their homes. Rheumatism and malaria were the two most common illnesses followed by sleeping sickness once they entered tsetse country.

Three men who had been carrying light loads for some time suffered from stiffness in the shoulders after carrying William in a *chupoya* (bed). They were cured of that ailment by being given heavy loads at more regular intervals.

Only the genuinely sick were excused from carrying loads and men who got lost in the bush were not treated sympathetically because they were suspected of either going too slowly or pretending to have been lost so that they could get a few days off on full pay.

When William and his men finally parted there were many expressions of regret and he felt it necessary to warn them that they must not publicly say that the English were better people to work for than the Portuguese.

6

From Catumbella to the Uaque River

William left Catumbella at 9.00 a.m. and was surprised when the lorry turned off along the Egito road because Mills had led him to understand that Blackwell's camp was along the road they had originally taken when they did a motor trip of the concession together. The lorry driver knew better and dropped William off about three miles south of Egito and quite near the sea. The place stank of meat but William was evidently expected because a clearing had been made for his tent. Blackwell turned up for lunch and gave William a warm welcome, saying he had been expected for the last three weeks but when he realized that Mills had made a mess of his arrangements for them to join up again, he did not hold back his work as much as he might have done. He had advanced a couple of camps and was now ready to move again. This had been a great camp for shooting and he had fed his gang meat almost every day.

10 June to 16 August 1930

Tuesday 10 June 1930
Soon they sent the lorry off with a guide and plenty of fresh meat to pick up as many of William's men as they could find and direct them to the right camp instead of letting them go off along the road they had been instructed to take.

The guide was then supposed to go to the other road to locate the advance party that carried most of the food, but they had gone off into the blue and could not be traced. They also sent almost a whole kudu carcass to Lobito to be distributed among the English population there.

Wednesday 11 June
This was a lazy day for William but in the evening he shot a klipspringer at about 130 yards and Blackwell shot a small buck. Blackwell was temporarily in charge of MacKay's dog and a small baboon. The dog, which was suffering from a scorpion sting in the throat, followed Blackwell out of camp for a short way but did not return. Blackwell took no notice of this, for it was not unusual and he was sure the dog would come back in a few minutes. This time was different and when it was dark Blackwell went out with a light to search for him, thinking that because the dog was ill it might be lying down to rest nearby. The following morning a search party eventually found a single paw within 200 yards of the camp, suggesting a leopard had probably eaten it almost under Blackwell's nose.

Thursday 12 June
A camp was prepared ahead and the morning devoted to Blackwell's plans. Then the lorry turned up with an irritating note from Mills. William was so enraged that he did no work for two days and soon found that two of the others sometimes went for weeks on end without taking any interest in their work.

Friday 13 June
The prepared camp turned out to be only half a mile from Egito so they went on beyond the town, much to the disgust of the carriers who had hoped for an easy day. The new camp was about seven-and-a-half miles (12 kilometres) away. It was hard for the gang to find material to make their *shengies* (huts), so most of them slept out in the open for the next few nights. Scouts had again been sent out to

find William's advance party only to be told they had been seen 'walking round various parts of Angola'. This was sad news for Blackwell because they had the food and whisky and he was now out of whisky and many items of food were almost finished.

In the afternoon William and Blackwell went to see the local *chef de poste* who was away at the time. They bought some *fuba*, barrels and other necessities for the two gangs.

As they were so near town Blackwell decided to pay his gang and soon they had all gone off for a drink and to spend their money. Some time after going to bed Blackwell and William were called out from their tents by Blackwell's *capataz* saying, 'There is a fight going on and one man has had his leg cut right off.' William slipped a pistol into his pocket and the two set off to the scene and found one man in a very bad way. His leg below the knee was only connected by a little skin and scraps of flesh. The medicine chest was totally inadequate for an emergency such as this. William tore up one of his shirts and a pair of shorts; he then broke up a petrol case and, padding it with grass, allowed Blackwell to make a job of sorts to reset the leg. This was a most unpleasant task, which he had to do in the dark with only electric torches to help. The drunk who was responsible for inflicting the injury was securely tied up and the *capataz* told he was personally responsible to see that the man appeared when required next day. This was one occasion when they regretted not having any whisky left to calm them down after such a trauma.

Saturday 14 June

In the morning William and Blackwell went to see the patient and found they had done a good job and that the main artery appeared not to have been severed. The numbness had worn off and the man was in considerable pain. They went round again to try and find the *chef de poste* who was still absent, so they reported the incident to his assistant. After hearing what had happened, the assistant decided that 'Since it was only a black man and it occurred while they were drunk, it is not worth bothering any more about the case.' Eventually, he agreed to go out for a drink with William and Blackwell and they were able to take him to see the sick man. By this time the man under detention had bribed his victim with all his wealth and promises of more to come, to change his story: he now said that while walking along the road he had tripped over a stick and in so doing had broken his leg. No one believed him and the culprit was taken to prison, nominally for trial.

The Portuguese declined to have anything to do with the sick man and the only thing left was to arrange for him to go to the hospital in Benguela. He was refused admission there and taken to Lobito for Mills to take charge. He was finally taken back to Benguela where, after further delays, his leg was amputated. The bill for transport was £5 and the man must have suffered greatly during the journey. Hospital expenses were very high and the question of who had to pay them was still being fought out in October. Meantime the culprit in prison had been sentenced and sent to another part of the country. His chances of ever regaining his freedom were considered to be very low.

William heard about his missing gang when they were about 18½ miles away and he immediately sent them a large ration of meat. Soon after the Portuguese official's visit to the camp they eventually turned up with all the food intact. They told him that for several days they had been without food and had become too weak to go any further.

The Portuguese inhabitants of Egito got to hear that there was whisky in camp and, as a consequence, they became very friendly. However, the supplies available did not meet their hopes, so one small group pressed William and Blackwell to join them for lunch at the local hotel. The best William and Blackwell felt they could do was to pay for the wine, which they did, but two days later they got the bill for all the lunches plus the wine. Later contact with Gordo, the

22. River Bolombo passing through limescale near Egito.

group's leader, did nothing to further their appreciation of his humour.

Monday 16 June

After making many enquiries about their chances of crossing the Bolombo River above Egito, they moved along a road on the south side. After two-and-a-half hours' walking they reached the reported bridge, only to find that it had broken down two years earlier and all that remained was a single wire cable and the piers on either side. The local farmer ran a ferry service but the man in charge declined to let the party use it without permission. It was not worth forcing the issue because the rest of the gang had to follow and could have been stopped. They therefore camped near the bridge and devoted a day to prospecting in that area. Tsetse fly was the only thing found, and it was the first time it had appeared since William's trip with Mills.

Tuesday 17 June

The necessary permit was obtained and the river crossed. It was a slow and laborious procedure and although it was supposed to be infested with crocodiles they did not see any. They moved on northwards for two-and-a-half miles before making camp. Here they did two days of work in difficult country with much thorn bush about.

Wednesday 18 June
William and Blackwell carried on work in the morning and, to their utter amazement and disgust, Mills appeared in the afternoon with three naval officers who had come out on a shooting trip. He greeted them by saying that he did not want to hear a single word of 'shop' spoken as he was out for a holiday. William did not like the captain, his main guest, who was very loud, but his one redeeming feature was that on several occasions he called Mills a liar, especially when he was describing his experiences of shooting elephant. They went out that evening and again the next morning, each time taking along a number of men to carry water and the game they were going to shoot, but they got nothing.

As soon as Mills and his party had left, William and Blackwell decided to take their lunch out and not return to camp until dusk. They had a very pleasant day in which they marked out the site for the next camp and noted that there were many signs of game about. On the way back they each took opposite sides of a deep ravine. Blackwell saw a buck in William's direction and waved frantic signals for William to climb down and along a very steep slope, most of the time quite unable to shoot even if he had seen the animal, but before he gave up in disgust his hands had been badly cut by thorn and cactus grass.

On their arrival back at camp they were surprised to see no light on and, to their delight, found their unwanted visitors had left that afternoon after another barren day. They had left a note saying that they wanted carriers sent after them to keep them supplied with water. As William and Blackwell had returned late the visitors were on very short supplies for the first 24 hours and relied on a basin in which to wash. Meanwhile, William and Blackwell had to make do with a reduced number of carriers who, because they were leaving water, were tied up for a couple of days.

Thursday 19 June
The glorious 19th was a suitable occasion to celebrate Mills's departure and the two sat up late with drinks and the gramophone – a luxury that was rarely possible.

Friday 20 to Sunday 22 June
These three days were used to prepare the next camp and get the water forward.

Monday 23 June
They moved to camp 36 only to find the men had pitched the camp site in the wrong place, about one-and-a-quarter miles too far, but to their credit they found a good spot. When they settled down in the evening it became obvious that Blackwell was even more fed up than William was about Mills's shooting trip and the loss of his men. He then admitted that he had already written to other companies to see if there was a chance of getting a job there, and all the others of the party (except William) had done so too.

Mills had made it very difficult for them to manage. They did so by keeping only six carriers in the camp; the rest had been sent off to sleep nearer water to ensure that supplies could reach camp 36 earlier in the mornings. Even with these arrangements in hand they were rather short until the remainder of the gang returned. The river was too far away to do more than one load a day and, in addition, there were many loads left behind that had to be brought up after the actual move.

Tuesday 24 June
An early start enabled much ground to be covered during the day, moving west into the limestone hills and almost to the sea. The country was generally very open, only rarely covered with belts of thick thorn bushes and small woods. They saw 17 kudu but they were not in shooting range.

Wednesday 25 June
This time the party moved in an easterly direction where the ground was covered in very nasty bush. Blackwell shot a kudu in the afternoon. On the way back to camp he started many grass fires to try and make the going easier as the wind was in a favourable direction. That evening William and Blackwell went out in different directions on a shooting expedition. William missed a bushbuck at full gallop at 25 yards but with a lethal ball hit a kudu, which the dogs helped him track down the next morning. By the evening William was suffering from a touch of the sun and a heavy cold, which to his annoyance took a week to get rid of.

Friday 27 June
Work was carried out in the plain and quite near the camp William saw three female kudu and got a good running shot at one of them. They followed the blood spoor expecting to find the dead animal very quickly. The trail became increasingly faint and after a two-hour search it became too dark to continue so they returned to camp. The animals were not always shot in an instant kill and William was now progressing in his skills of casting forward when the spoor was lost and in reading tracks.

All the carriers were back now including those Mills had taken for his shooting party and the messengers sent to Lobito. The carriers were very disgruntled because they had had no meat to eat. There had been plenty of shooting at animals but they only brought down one kudu, which they took back to Lobito. The carriers thought many others were wounded but practically no attempt was made to follow them as the wounds appeared to be slight.

Saturday 28 June
The men went out early to find the kudu, which was standing in a stupor and unable to move. Blackwell's shot broke a leg and brought it down and William finished it off with a lethal ball that penetrated right through both

shoulders, coming out on the opposite side. Shooting kudu, bushbuck, klipspringer, steenbok and another small buck provided a variety of meat including liver, heart, kidneys, tongue and the cut from the centre of the back, which is the best part of the meat on all game. Birds were plentiful but almost unwanted except for occasional additional variety.

Cash was short and Mills was unable to give them any, and the whisky was again almost finished. Other stores that had been ordered also failed to arrive and once again William and Blackwell vowed never again to trust the company representative. They would have to purchase from the Portuguese who, although they charged more than the British firms, did at least send out the items requested. The main food shortage was fruit and vegetables, which at times became serious.

The mail was another problem and one package had been sitting in Mills's office for almost two months. He had previously held up a package of Blackwell's for two months and other letters and parcels were delayed unnecessarily. As letters to and from home went astray the fault was probably at Mills's office.

Sunday 29 June

Having collected three kudu in as many days the camp was moved a short distance along the road to Novo Redondo. This was a difficult move because each man had to shift three loads carrying water and the extra meat supply. Much of the meat was cut up and dried for biltong. As the track became more worn the journey time decreased until they saved over half an hour taking about two-and-a-half hours each way.

Jacko, the monkey, usually followed like a dog and whether on a march or out prospecting in the bush he always kept close, picking up odd scraps to eat and always very alert for any other animals. He was terrified when guns were fired but was a useful 'pointer' in that he was frequently the first to spot game, at which point he would rush to the other side of William or Blackwell away from danger. While walking he would climb every stump or vantage point to have a look round and let out various cries if he saw danger, which never disturbed the game and appeared to attract other baboons. It was usually easy to catch him within the last half mile to the camp.

One evening Jacko surprised everybody by stealing Blackwell's belongings in the dark when he could not be caught. MacKay normally kept him loose, but when he wanted to tie him up he was sometimes difficult to catch. If all else failed he would get the dogs to chase him, whereupon the monkey would either run up a tree or bolt straight to his master for protection. The dogs knew the game and enjoyed it immensely. Once up a tree a number of men would pelt him with sticks and stones until he dropped to the ground in fright and the dogs would have another chase after him until he ended up with MacKay or Blackwell for protection and could then be tied up.

Jacko was particularly fond of stealing bread, papaws, biscuits and jam. He was adept at opening the biscuit tin and was apt to be destructive when on top of a tent. He stole several things from William's tent but knew he had to be careful because the men had permission to throw what they liked at him whenever he went near William's belongings. William felt it was time Jacko was taught a lesson. As he ran to safety from Blackwell's tent he dropped a small loaf of bread (which was in very short supply at the time). William picked up the loaf, cut it open and spread his delicious jam inside, seasoning it with a large salt spoon full of red pepper. The sandwich was neatly closed and put in a biscuit tin, which was left in a conspicuous place. Soon Jacko was there and went straight to the tin, chuckling almost as much as his onlookers did when he found the bread and jam. He quickly moved to comparative safety under a bed, from which he could easily run, and there he started his feast. After the first mouthful of jam he stopped to have a good look at it. He was suspicious for a long time and gave the jam a miss, eating the bread all round it, but greed won and soon he went back to the jam and managed to eat most of it, finally only leaving just a little bread and jam. He spent the next half hour holding his nose against the metal part of the bed frame or anything else that might cool him. The onlookers had a good laugh and felt satisfied they had scored a victory. They retired to have a bath and change for dinner. Soon William heard an agonized cry from Blackwell. The table had been laid for him and as he was about to sit down to dinner he discovered that Jacko had got there first and had taken his last loaf of bread!

Monday 30 June
It was not difficult to catch Jacko in the morning because leopards had been round the camp all night and he was very frightened.

The next few nights were very cold and raw and the water from the morning mists running off the tents was as though it had been raining. Most of Blackwell's belongings were soaking wet but William had his curtains up and, although they were wet, they kept the moisture out of his tent so that he had dry clothes to wear in the morning.

They were far away from water and as a move to the River Eval was reported to be too far to march on foot, they hired a lorry. They assumed it would be at least two hours' late, so when it came only half an hour late they were totally unprepared. They had not yet taken down the tents or packed the meat, partly because they wanted everything to dry out as much as possible before starting. The journey to the Eval River turned out to be only nine miles (fifteen kilometres), which William thought they could have marched had they been allowed to borrow the company lorry for one day when it was idle while Mills was holding his shooting party. This way the two gangs would not have been held up for nearly a week because of the water shortage and the exorbitant hire price could have been avoided.

23. Crossing River Eval on coast road in the dry season.

On hearing the price charged for the hired vehicle they unhesitatingly loaded it up well above maximum load and managed to get everything moved under the one charge.

It was a bad day when nearly everything seemed to go wrong and the carriers were slow and unhappy. The river turned out to be a small shallow stream about ten yards wide and safe from crocodiles. There was plenty of fresh spoor from new game including bushcow and roan antelope. Leopards were prevalent and a herd of elephant had passed there the previous night. Consequently, a number of the townsmen wanted to be paid off to go home, for they had had enough of the dangers of camp life. It was impossible to replace them and as they had had a chance to leave when they were passing Catumbella, William felt that he was not being too unkind by keeping them on for a bit.

Tuesday 1 July

The party moved downstream to the schist–dolomite contact and made a bad camp on high ground in thick bush. Such was the terrain that it was hard to clear even two level bits of ground on which to pitch tents, and the tent pegs were badly broken on the rocks.

On the march Blackwell shot what he thought was a sable antelope. It was badly wounded in the hind quarters but managed to get away. His men tried to track it for three days until carrion crows attracted their attention and they found it had been killed and eaten by a leopard. Only the head remained, which they brought in for Blackwell to keep as a trophy.

Wednesday 2 July

While out in the early morning in patchy mist an unusually large bushbuck was spotted. The animal was suspicious but after stalking closer to it William got into position for a steady head shot. The two dogs went racing off yapping. On reaching the spot where the animal was last seen it was evident it had dropped stone dead where it was. As there was a morning's work to do the animal was hoisted up a tree to a safe height for collection later. After work was done it was difficult to find the tree again but the men were not going home without the evening meal. It weighed over ten stone and was an awkward load to carry in difficult country.

Three days work was needed to find and prepare the next camp to the north of the contact. The bush was very thick and the banks of the River Eval rose steeply to about 500 feet, which was too difficult for carriers to get up with loads.

Sunday 6 July
Before moving across the river William found some interesting country downstream on the limestone–gypsum contact. On Sunday they moved there with minimum loads, making a nice camp in a dry creek bed. In the afternoon William had a pleasant stroll that ended in a heavy climb and he finally got a small kudu bull. The descent down the limestone was unpleasant and painful, with the result that William and his man did not reach camp until long after dark, just as Blackwell was about to send out a search party.

Monday 7 July
The camp was moved further upstream and they settled in just after dark, but a new set of troubles greeted them. Messengers had been sent out to try and locate the other parties near Seles but they could not be found, while other messengers had been sent to Lobito for urgently needed cash and whisky. These had been given a lift back by their 'friend' Gordo (the man who had stung them for lunch and drinks in Egito). When the goods arrived they were two bottles short of whisky and two bottles were broken. On further investigation it became obvious that the broken bottles had been consumed before the necks were hit with a hammer. There was no stain on the wrapping paper and the bottles could not have received such blows in the manner in

which they had been packed. Some money too was missing. Gordo had taken as much as he required for himself and made various excuses for his accounting practices. They were too far away to deal with Gordo personally so they sent a messenger to get more cash and hoped to recover the money from the next bill due to him. By this time he had sold his account to a friend in Benguela and Mills paid it without even checking to see if it were correct. They were all out of pocket by about £4.

Tuesday 8 July
William moved to camp 43 upstream near the road to try and intercept Gordo, but without success. It was a bad move because most of the spare stores, kit and food had been sent to the prepared camp north of the river and all this had to be collected again. There was also a fair supply of meat to carry and William found the stench walking at the tail end of the column so abominable that he quickly moved nearer to the front. After the camp was set up he marked out the site for the following day.

Wednesday 9 July
The men carried out the move across the river to the camp on the northern bank without supervision. It was well hidden but near the road. By now they were pretty scared because lions were evident as well as numerous leopards. The smell of the meat attracted the leopards, which persistently stayed near the camp at night.

In the morning a herd of roan antelopes was seen about a mile away. After breakfast William and Blackwell went in search of the animals, soon picked up their tracks and followed them with difficulty for a long time as they wandered along feeding. They spotted a big bull and while Blackwell covered the animal William was going to shoot it. William stalked it and got into a good position, but the animal would keep twisting round and moving out of sight. He looked magnificent, the size of a dray horse, with fine sloping horns and in magnificent condition, moving with the speed of a thoroughbred, so fine that William was reluctant to pull the trigger. In the long dry grass it was hard to follow the spoor. After a long search the last view of him was staggering away in the bush, just too fast to be followed. Two days later he was found stone dead. He weighed more than a thousand pounds and his meat was dried for biltong.

All possible carriers were sent to Egito for food and this left the camp too short-handed for work for the next five days. During this time William selected the next camp near the foot of a big hill alleged to contain much copper but it was ten-and-a-half miles (seventeen kilometres) from water. It looked like a good spooring camp with plenty of bushcow spoor about, but a local hunter advised that no man should go near the area on account of the number of bushcow and an equal number of lion about.

Monday 14 July

Advance loads and water were sent to the 'danger' camp while various men tramped off independently to try and find fresh fruit or vegetables from farms up the river. The farms were all deserted and there was no sign of any fruit anywhere.

Tuesday 15 July

In the morning they walked for three hours along a good road and then turned off for a mile. In the afternoon a preliminary walk round caused their legs to be badly cut by thorns hidden in the long grass. As the wind was favourable they started a number of fires, which made it possible to move much closer to the mountain to be prospected. To make things still easier they first prospected in easy country to the east, setting both gangs, minus the water carriers, onto starting more fires. The fires burnt for nearly the whole night before dying out.

The water problem also appeared to be improving, with

waterholes being reported by local men who had just been engaged. As it was long after the rains it was anticipated that this information might be incorrect.

While William's gang moved his camp, Blackwell carried water for both gangs. William went to see the waterholes, but found that the total supply barely reached a barrel per hole and that the distance was almost equal to going back to the River Eval, where at least there was running water.

Friday 18 July

Blackwell was suffering from the lack of vegetables but refused to try William's remedy of maté. The days were hot and their preliminary prospect along the foot of the mountain was hard work, nearly all the distance covered being under bush, continually stooping and cutting to get forward. No game was seen throughout.

William took a long walk in the evening and returned to camp just after dusk to find that Mills had paid a visit, which was expected but not wanted. There was insufficient water for three white men so William left a note for him at the roadside to tell him that he would be short of water unless he brought his own supply in the lorry. He failed to do so and was bluntly told there was not enough for him to have a bath. William and Blackwell each had a bath while Mills's man got less than half what he wanted for the evening meal and for washing up.

That night William found a letter in his tent from Mills. William had asked the Angola Coaling Company to get him some whisky from England at a special rate. There was the usual delay to get the consignment through customs and each time he was in touch with Lobito for anything he would enquire about his whisky. Eventually, he received notice that it had been cleared. Mills took the case (as he had done the previous month) falsely claiming that he had cleared it with William. Mills did not tell William that he had the whisky, offer to pay for it or say thank you. Three messengers had returned since he took the whisky, giving him ample time to deliver it or at least offer the minimum courtesies. William wrote disassociating himself entirely from the deal. He replied enclosing a cheque for too large a sum, which William returned with a note to say that he had better complete his deal with the Angola Coaling Company directly, and that they were not charging as much as he seemed to think. Having drawn him sufficiently William finally accepted his cheque for the right amount. He later confessed to Blackwell that he had been in the wrong.

Saturday 19 to Sunday 20 July
Mills toured part of the countryside with Blackwell. The first morning he went to the mountain along a moderately good path. When he reached a point a short way up the hillside he proposed sending the men with him to collect rocks from high up. They flatly refused to leave him, saying that it was bad country and that the local inhabitants would kill them and chop them up. Mills then conducted his prospecting in more open country and to a considerable extent from the lorry.

William went off on his own on Sunday starting out to the west and, as he circled round to the north, he found water barely two hours away from camp. Back at camp he told Mills he planned to move his gang there next morning to which Mills readily agreed, suggesting he covered a belt towards the coast.

Monday 21 July
William set off at a decent hour with some of Blackwell's carriers and under Mills's very disapproving eye. The latter two went off in the lorry to make plans for a further move northwards and to check out the water supplies – and probably to seek out some better shooting.

Tuesday 22 July
Following a note from Blackwell William went to see him for a chat. Mills had given impossible orders that had to be sorted out. William was to go to the Uaque River, which

only had a feeble trickle of water, while Blackwell was to go east and then they were to meet up and work north along the river, each taking one bank. The main problem was that the river did not go north, but south. His information was based on Portuguese sources received from a local through various interpreters. The two decided that William would work from the sea upstream while Blackwell, using as many of William's carriers as he needed, would work as a flying column going upstream to the granite, round a volcanic contact and down to join William as soon as possible. Meantime, William would move all his unwanted stores and food down to the dolomite–schist contact, and make a central camp there.

Shortly before William started back for camp a Portuguese man escorted by a number of others came up to the camp, which was now near the road. He lived in a farm nearby with four Portuguese men and many local workmen. They had used up their last round of ammunition and this lad had been sent to Blackwell to borrow firearms with which to protect themselves. They asked for his rifle and he reluctantly gave them his shotgun and 25 rounds of ammunition. They told him how foolhardy he was to live the life he did and that they could not understand how he had lasted so long in such dangerous country as they had just passed. Blackwell did not reveal that they had not encountered a single bushcow, lion or more than a few local people, none of whom seemed troublesome. There were a number of leopards around the camp at night but they never became a menace.

On meeting up again Blackwell accompanied William part of the way back to his camp after which they parted company for several weeks.

In conversation with Mills, Blackwell sang William's praises. He showed him the magnificent heads (Mills had yet to get his first) and all the meat hanging up to dry. He explained that not only did they have no reason to buy fish or meat for months, but at times they had been able to reduce the *fuba* rations as well. Although sickened by these praises Mills admitted that their gangs were the only ones covering a decent amount of ground, although they were rather short of carriers. The others were quite immobile for want of decent men.

The Uaque River prospect: copper at last

Thursday 24 July
William had an easy day and only went out for two hours as most of his carriers were taking water to Blackwell's flying column and a few were looking for a route downstream and reported getting a long way down.

Friday 25 July
William set off along the route going downstream, taking some loads in advance to make a camp on the south bank. The going was bad, the hills being small but sharp and frequently covered in dense bush that required much cutting. After struggling for two hours William decided to try the north bank but it took a further hour-and-a-half just to cut across the stream bed, which was covered in dense reeds about 15 feet high and then scramble up the slope on the far side. Soon he selected a camp site, built a single *shengie* for the baggage and returned to camp 46. The walk out took four hours with baggage; the return was completed in an hour-and-a-half. On the way back a fine bull roan antelope was on the path ahead but William's gun carrier was too far behind to get it in time. Fresh meat was wanted but there was still plenty of biltong hanging up.

Saturday 26 July
Each man had two loads to bring up and camp was ready by 2.30 p.m. Having seen quite a lot of fresh spoor William went out after lunch equipped with his rifle, glasses and his haircutting paraphernalia to take up a vantage point and look out for game and perform the necessary operation on

his hair. The idea of cutting his hair was soon dropped in favour of a roan antelope browsing under a tree some distance away. The stalk was easy but long, for he could keep to the opposite side of the hill walking freely and then approach with the wind in the right direction. He had to be careful crossing the crest but soon found his quarry was a huge eland bull, which was moving slowly away. As he followed it fairly fast he suddenly came across another one feeding to his left front. It was smaller but an easier shot and would probably give enough meat to last the gang a fortnight. He shot it in the heart from a range of just over 100 yards, but he went off at a gallop apparently unhurt. The big one sounded like an avalanche as he galloped off. Being short of ammunition, William rationed himself to two rounds. He caught up with the wounded one still standing in tall grass, but the second shot went over the animal, which bolted a short way and then started a painful walk. When William got close to the spoor he was sure the eland would die soon. He went back to camp for his shotgun and lethal ball and to get carriers to cut up the carcass and bring it back to camp.

They got back an hour-and-a-half later and soon spotted the bull, but although much the worse for wear it was still able to get away. William followed it alone and when he got near enough he shot it again with a lethal ball, which made it stagger but it got back into thick bush before falling. Another search and from short range another shot went to its heart, but again it struggled to its feet and got away for about 50 yards before dropping. This time all the carriers and their dogs rushed up to finish it off but it was not ready for that and in a moment had cleared a respectful circle for itself. As the lethal ball was expended the last remaining weapon was a .32 pistol with eight rounds. At short range William fired three shots to the forehead, three behind the ear and one just below the eye (all trying for the brain) and a final one behind the shoulder for the heart. They seemed to have no effect whatever. Further attempts to lasso it to turn it on to its back were met with a struggle to its feet and a charge. It was getting dark and the shooting party returned to camp with the eland having won the day.

Sunday 27 July
In the morning they returned to collect the eland, which had again struggled a short way from where they had left it. It provided 20 big loads of meat and bone and altogether must have weighed over 2000 pounds. In the postmortem there was nothing wrong with the first shot, which had hit it in the right place and had broken up well, one fragment making a jagged wound at least 15 inches deep and another fragment was actually in the heart, yet still it could gallop. In addition, the first lethal ball had broken a hind leg and the second had smashed the shoulder blade and part of the wall of its chest. The pistol bullets to the head all failed to get through the bone and gristle, but the final one ended up in the muscles surrounding the heart. It made William gasp to contemplate the fight left in the animal after all those wounds.

The eland is a rather shy and harmless animal that keeps away from humans as far as possible. They are rarely encountered in herds of more than a dozen and are most often seen in pairs and sometimes singly. In herds the bulls are always out at the corners grazing, keeping a lookout for danger. They move at a slow walk but if disturbed may break into a steady trot for a mile or so. They very rarely gallop, even after being fired at. The eland has no teeth on the top front part of its mouth. The palate is covered with serrations, which, with the tongue, are sufficient for grazing. They are said to prefer browsing on trees to eating from the ground. Their horns reach a maximum length when three years old after which they get increasingly shorter.

William's later personal experience at camp 55 was that the eland was inclined to charge. It was here that he spent much time at dusk trying to get close enough for a shot. As it became darker so William had to get nearer the animal until at length he had to give up. The following morning he trailed the animal again only to find it had turned on its tracks and had stood facing the spot where William had stood apparently ready to charge. At such short range it would have been impossible to stop it.

Back at camp William found that Blackwell had taken all the salt and so all this fresh meat had to be hung for biltong without preservation. A *shengie* was hastily built to keep it cool by day and it was hung in the open during the night. Meanwhile, messengers were urgently sent with a whole leg to Blackwell and a plea for salt to be sent back as quickly as possible – a bonus was offered to the men if they got it back before the untreated meat went rotten, which they did. For some time after this there were many yards of meat hanging outside William's tent and, with such good meat rations, the men were particularly cheerful.

Monday 28 July
William sent forward all available carriers with advance loads to make a camp about three hours ahead. They returned rather early and reported that there was much thick bush ahead but for the return they had found a much better path.

Tuesday 29 July
William accompanied his men and moved camp only a short distance and from there explored a better route before going on.

Camp 48 was to be a depot camp with two men left permanently in charge of it. All the food, stores and unwanted loads would be kept there until William returned to rejoin Blackwell. The meat had not yet dried and could not be safely left with the two men, so the next day he went on with it and some *fuba* to prepare another camp. For over two hours the going required a great deal of cutting and the cacti and thorns were very annoying. The site was a bad one but the return route was much better and only involved half an hour of bad ground and climbing to overcome, but even that was avoided when shifting camp on the 31st. By now another messenger had come from Blackwell bearing bread, wine and papaws and leaving William with few wants in the world.

Thursday 31 July
This camp was on the schist–dolomite contact again, and in the morning William sent all the most intelligent carriers out in various directions to collect rocks and to look out for copper, and in the afternoon he went to inspect the most promising area. The bush was still very thick and most of the climbing was severe. The *capataz* and some of the men meanwhile found another track forward.

Saturday 2 August
This track was followed to reach the junction of the Uaque and Eval rivers after crossing the dolomite hills, which were about four miles wide and were covered in exceptionally thick bush. Luckily, the path was in regular use and only a little overhead cutting was necessary to make it easier for the loads to get through. There were no difficulties other than trying to find a good camp site, which turned out to be only two hours' walk away and so allowed second loads to be brought up the same day.

Monday 4 August
This was a bank holiday but for William it was another day's work. He moved camp about five miles (eight kilometres) north in the hope of finding another route back by keeping close to the bed of the Uaque River. From the moment the river left the schist country it ran entirely underground, only rarely emerging in the form of a stagnant pool. The vegetation along the bed was exceptionally thick in many places.

The track went on in an open valley where practically no obstacles were encountered. Indeed, the route was so easy that William sent the carriers on without him to set up the new camp while he took his personal servants, Macai and Kalcoulete, on a mountaineering expedition with a table and chair to fill in the district on his map from the various high points in the neighbourhood.

It had been an easy day until they attempted to descend on the side near the new camp. The grass was barely two feet high and not dense but the gradient was steep. The northern slopes were comparatively easy until they reached the bottom 500 feet when, in addition to the slope, they came across thick cactus grass, a couple of ravines that proved hard to negotiate and, eventually, an unscalable cliff.

Macai was out largely as a punishment because he had been showing signs of growing slackness, and this walk proved an admirable one to cure him. He had got pretty soft and, in addition to perspiring heavily and getting tired, he had grown unaccustomed to walking through cactus grass without pain. William and Kalcoulete derived sadistic pleasure from watching him put his jacket on each sharp point that he had to pass and almost invariably, in his attempt to avoid one spike, pricking himself badly on another. Their fun soon ended when they had to retrace their steps back to the stream bed, which was heavily overgrown and they could find no way through. They had no bush knife with them and cutting with William's hunting knife was slow and laborious – it took two hours to cut through and clamber over 30 yards. Another half hour's walk brought them back to the camp to which most of the carriers had returned with their second loads.

Macai's ordeal was not over yet. He had to get William's lunch served before he was allowed to sit down and rest for the remainder of the day. He made no attempt to put on his best clothes to go and see his mostly female friends at the neighbouring farm.

William went out along another dry creek bed to the north in the afternoon where he encountered much thick bush until it could no longer be penetrated. He was lured forward because the area looked promising.

Tuesday 5 August
After much pitting William found there was insufficient copper to be of commercial interest.

On Monday and Tuesday the scouts had failed to find a satisfactory route along the river bed, so this afternoon William broke camp and went back to camp 49, which was rather a long walk. On the way he was met by most of Blackwell's gang, who had a note from him saying that he was unwell and had gone to Seles with Mills to help with the monthly report. Since these additional men put William's team well above the requisite number, he sent them back to camp 51 to collect the rest of the loads. They made a good start by deciding they had walked far enough that day and did not appear again until the next morning. Another problem was that they came with no food and William barely had enough for his own gang.

Wednesday 6 August
While waiting for Blackwell's gang to collect the loads William went out into thick bush country and, coming to a barren area of malachite, he found a small trace of copper. This was the first sign of metal he had seen since leaving Mount Assango on the Cavaco.

Thursday 7 August
As soon as Blackwell's gang had brought the loads up they had to go off to get some food. William made camp a long way from water as he had a full gang and could afford to have a number on water-carrying duties. The move was again in difficult country where much cutting was necessary. The bush then opened out into an area where shrubs were far apart and local hunters had burnt off the grass.

That evening William found two of Blackwell's gang still sitting in the camp. He fined them and told them to leave the camp at once. He could not sack them because they were not his men, but this was sufficient to have an excellent effect on the remainder. After that he gave his orders for the following day's work.

Friday 8 August
At dawn a herd of roan antelopes was reported near the camp and William immediately set off after them with his gun, but having not got near enough after two-and-a-half hours he returned for breakfast. The rest of the day he spent with the prospecting pick and mapping.

Blackwell's gang failed to return until Saturday, which was a blessing because there was insufficient water for 20 more men.

William was disturbed by a fight in the camp after going to bed. Apparently, his servant had taken his flour sifter into his hut where it got broken and he was anxious not to take the blame. His antagonist speedily cleared out of the camp, never to return again, and the culprit had to pay for the damage.

When William recovered the sacks of dried meat from the depot camp some had gone down considerably in weight and the two men in charge had to pay for the amount they had consumed. In addition to the shortage of meat there was no oil, tobacco or salt in Blackwell's camp, all of which were listed on his inventory.

Saturday 9 August
William found an area with a certain amount of malachite between camp 53 and the river, and Blackwell had also found some before leaving for Seles.

Monday 11 August
After the water carriers and a party to accompany William on his own general prospect had been sent off, the rest of the men were busy sinking pits on these two copper finds to see if they had any depth. When William returned to camp at about 3.00 p.m. he found an urgent note from Mills saying that he was waiting for carriers on the roadside and

that they were to be dispatched to him immediately. As usual, he was not travelling light and to meet his requirements William had to move his camp nearer to water as quickly as possible. William sent the whole of Blackwell's gang and would send the balance of his gang next morning. The first of Blackwell's gang were soon on their way.

After lunch William went to see the rest of the gang that was pitting on Blackwell's copper find. Unfortunately, he missed them on their return route through the bush and only met them back in camp at dusk. That night they moved to the store camp with orders to move on as soon as possible the next day. Mills had forgotten how far William's camp was from the road and, having the lorry with them, he and Blackwell had a free day to go shooting.

Tuesday 12 August
William moved camp to near his copper find and only a short distance from water. Pitting continued and he also completed a three-hour prospect. By now William had very sore and swollen shins that were discharging considerably. During the day there were exchanges of notes because William wanted Mills to see the find and to turn it down as not worth further attention. He agreed to come out the next day but later made excuses not to go.

Wednesday 13 August
Blackwell arrived with a letter from Mills giving details of forward plans that entailed the two gangs being separated. It therefore became necessary to join camps once more to sort out the equipment. William sent some advance loads to Blackwell's camp and was happy to entertain him to lunch because his feet were still very sore. Blackwell was unimpressed with the only water William had to drink, which was not clean. He was also short of many minor articles of comfort, which a messenger was sent to fetch from his camp. With this help Blackwell supplied the meat and vegetables and William produced the cook and the dessert.

Thursday 14 August

Macai returned to the last camp to collect a second load and had hoped to get someone to carry it for him, but was unsuccessful and had to do it himself. He told William that he had seen some spoor on the way out, so William went to look for it but saw nothing under a week old. By the time he had walked on to Blackwell's camp one leg had become very swollen and painful from veld sores. Fortunately, it was a problem that Blackwell had suffered and he had a small supply of mercurial ointment, which they divided. This opened up the wounds and stopped further swelling. While caring for their wounds and dividing the stores, gossip inevitably followed. On hearing Blackwell boasting that he had hit three roan antelopes in one volley, William told him off for shooting more than he needed at the time. Blackwell bowed his head in shame, claiming that he had lost his head. Mills had also had a lot of shooting, his best effort being at two very young eland. Seven rounds were fired at gradually reducing range but no hit was observed. He finally advanced to close range and sent them galloping off by throwing stones at them. He then left instructions with Blackwell to purchase some fine heads from the local hunters!

A black mamba was disturbed by Lisa, Blackwell's dog. The snake went straight for the dog but missed it. It then went for Blackwell who saw it early in the burnt-off grass. His first shot with the shotgun missed but the second killed it at very close range. The snake was about three feet long and had recently swallowed a large lizard, which occupied nearly half its length and was as yet undigested. Blackwell was rather shaken by this episode and in the knowledge that if a man kills such a snake the mate is apt to follow the killer for a week or more, he was thankful that he had to return to Seles.

At camp 55 a number of men had given Blackwell the requisite notice and with their loss he was left very short of staff. Now five of William's men were feeling homesick and a sixth, who was ill, was also allowed to go. This left William short too and he allowed Macai to help Blackwell out because his personal staff had left. William's benevolence was due to his getting some much-needed bread and fruit, which had come free from the Portuguese in return for the shotgun that Blackwell had lent them. He had great difficulty getting his weapon back and, although the ammunition had not been used, they refused to return it. After that they had not the face to charge for the bread and papaws, which had cost them nothing anyway.

Saturday 16 August
Blackwell left camp with tears in his eyes as William sent advance loads to his copper deposit and had a camp site prepared there as it was close to the work in hand and the tsetse fly were becoming worse at this camp.

7

The Uaque to the Cubal River

William moved up to the copper area on Monday and found it to be a rather disappointing camp. He missed the shelter of the trees, but at least there were no tsetse fly around. The grass in the whole region had been burnt off, but the common fly was a nuisance from the start.

18 August to 10 October 1930

Monday 18 August 1930
William's leg was so bad he had to be carried. The make-shift *chupoya* was made from two old sacks, ropes made from the bark of a local tree and a few nails that came out of an old box used for petrol tins. The bed was supported on two long poles, which were long enough for the *chupoya* to support William's body but not his head. As he was very close under the pole he had to keep his head on one side or the other to avoid getting banged about any more than he was already. He was furthermore totally unprotected from the sun and quite unable to keep an eye on the country traversed. The path was too narrow for more than two men at a time and so two pairs took it in turns to carry him. For the first half hour all went well but long before the final climb was accomplished the carriers were exhausted and were more than ready to accept some outside assistance.

In camp William found that the various messengers sent out in 'hopeful' directions had returned with promises that more labour would be coming in and they brought with them a plentiful supply of bananas. William was so delighted that he ate 15 in one go and his appetite was not impaired when the cooked meal followed. During the short period that William's leg would not permit him to take sufficient exercise a normal meal consisted of a plate of soup, at least a dozen bananas, three or four oranges, followed by bread, bugs and cheese – the bugs generally thrown in with the bread gratis.

The new camp was on high open ground and was much cooler than the last one in the trees by the river. In the afternoons it was possible to enjoy a pleasant breeze coming up from the sea to the higher ground.

Tuesday 19 August

Because Blackwell had already covered the area round this camp in the morning William was carried to the north along the contact to select the site for the next camp. The River Cubal, the next water ahead, was said to be about 15½ miles (25 kilometres) along the schist–dolomite contact from the Uaque and, in view of the recent copper finds, the ground had to be gone over with extra care despite probable water difficulties.

Things were looking bright and the country ahead looked delightful with mile after mile of long rolling downs with not too much grass and little bush.

William took water, barrels, tools and a few loads of *fuba* to prepare the next camp. After going about two miles a roan antelope was seen feeding some distance ahead. All the carriers were told to lie down while a fast runner was sent back to get William's rifle. Meanwhile, William, forgetful of pain and probable consequences, set off to stalk the animal wearing the lightest of bedroom slippers. He soon became aware that his footwear was not the most suitable for the occasion, but keeping in a hollow out of sight of the animal, and walking where the grass fires had made it easy, he was eventually able to get into the desired position for a shot, by

24. Bananas, coffee and more.

which time he had got his rifle. At the critical moment he was guilty of greed and having covered one animal he spotted another bigger one, and the distraction enabled them both to get away. The carriers picked William up again and carried him until he could point to a suitable camp site. At that moment he spotted a herd of roan antelopes grazing a long way off. He was carried towards them and at a suitable moment he set out for another stalk. A couple of isolated eland bulls were also there but disappeared quickly. There were thirteen in the herd, including four good-sized bulls, but they spotted William and ran off. There was no chance of catching them and long before the carriers picked him up his leg was aching furiously. On returning to camp he spent the rest of the day treating the sores. Luckily, his attempts to shoot meat had done no further damage to his leg.

Wednesday 20 August

William did not have enough carriers to spare any to carry him, so he made a determined effort to cope. The distance to the new camp was not far and by walking slowly and carefully he managed it without further damage being done. The camp was sited in a small thicket on top of a long easy sloping hill in beautiful open country. It was the highest hill around giving a fine view of the wide-open spaces so often heard of but rarely seen. For most of the day William had ample time to rest his foot. Fruit and vegetables were at last obtainable, giving his blood a chance to get right again.

Thursday 21 August
The rest left him feeling well enough to start prospecting again in the afternoon. Messengers were at once sent to the road, a day's march to the east, to intercept Mills, who was supposed to be passing that way either today or tomorrow on his way to Lobito. William's men waited nearly a week before giving up on him. Apparently, he had gone to the coast three or four days earlier by another route.

The most recent problem was that William had only been given large banknotes, from 50 to 500 angolares, and when purchases were made often no change was given. The men were paid 50 or 60 angolares (10 shillings) a month and in many cases would lose the whole note when paying for a small article that cost only a quarter of the sum they had paid because they were given no change. They were too frightened of the Portuguese to complain and if they had they would probably have been told that no change was due. In the end William collected all the large notes he could and messengers took them to Lobito to exchange them for smaller denominations. It was almost a fortnight's walk there and back and fortunately the messengers did not know the value of the packets they were carrying.

Friday 22 August
William managed a moderately long walk at last and found some good game tracks to follow. He saw some game but considered it was not worth taking too many liberties until his leg had healed a bit more.

The carriers were busy doing two water trips a day to build up a good reserve for the next move forward. Today half the carriers under Mattheus were sent back to the farm on the road some miles from camp 45 for food, a trip that was expected to take two-and-a-half days in all.

Saturday 23 August

On returning to camp William found that half his supplies of dried meat had been stolen, so he pronounced that the carriers could have no fresh meat over the weekend and no tobacco for two weeks.

William moved forward to camp 58 along with the usual prospecting party, guns and as much water as possible. The next available supply of water would be at the River Cubal, but a man sent on to locate it reported that after walking for many hours he had still been unable to find the river.

The camp site turned out to be delightful; it was hidden in a small fold in the ground in slightly rolling country of easy grass without thorns and only lightly covered in small woods and light scrub. It was an impossible spot to find except by tracking the route already taken. Once the ground was cleared and the barrels filled all the men except for William's personal party were sent back to the Uaque River to get the day's water requirements, while William went forward, determined to see over the next horizon.

They were getting short of ammunition and since attempts to shoot birds and game over the last few days had been 'disgraceful' William decided to shoot only if he was certain of getting meat. He saw 16 roan antelopes feeding but, after watching them for five minutes, they went off at a full gallop. A few minutes later he came across another small herd sheltering from the sun under and around a cluster of bushes at the bottom of a small valley. William and his personal assistant scuttled back behind the crest of a hill without being seen, and then a short and careful stalk brought them as near as they could hope to get without being seen. It was a range of about 350 yards, much too far for accurate shooting with a rifle, but as it was impossible to get nearer William decided to have a go. He crawled under a bush where he could get a steady rest for the gun yet also remain hidden. Once in position he could not resist a shot and the first one dropped a cow standing in a group. It fell without moving and weighed 500 pounds. The remainder dashed out into the open and started to look out for the culprit but failed to locate his hiding place. Another shot hit the biggest bull hard but far back. He hobbled a few yards and then lay down dead, facing William in the shade of a bush. By now William only had one cartridge left and in a moment of greed went for the other big bull, but this one only flicked his tummy; it made him jump considerably but he seemed to be unharmed. The herd then slowly moved off up the rise beyond the valley. William followed under cover thinking he might get one more animal with a lethal ball from his shotgun. There was one bit of difficult ground to cross and he managed this, but his assistant was spotted and the herd went off at full gallop. At that moment William spotted what looked like a herd of cattle in another direction, which immediately suggested a farm and the availability of water. William got out his glasses but found the cattle to be a herd of 35 eland – a quite exceptional sight. He thought he saw a good approach route and hoped his shotgun would be adequate for the job. The herd started moving, but William saw a way of getting ahead of it without giving off his scent. Despite his care, a big bull that had hung back an unusually long way behind the herd saw him and went through them at a wild gallop; then they were all gone. William had gone far beyond the ridge he had aimed for and was annoyed that there was still no sign of any water, for the camp was three hours' away from a supply. They were getting near to having to consider giving up baths and washing up.

Sunday 24 August

William went out with the cook and two other men to cut

up the young roan antelope shot the day before. He sent the cook back with some meat to prepare for lunch while he and the others stayed another half hour to have a long walk round. On returning to camp there was no sign of the cook or lunch. As the cook was obviously lost William got out his whistle and made a dreadful noise frightening all the game for some distance around. Some time later the cook arrived back confessing to having been lost. He said that if he had not heard the whistle it would have taken him much longer to return. That evening William went out with his gun, roughly circling the camp about two miles out. He spotted a roan antelope out in the open but not far from a small shrub. William stalked up behind a shrub that was just out of range for a shot. The roan spotted there was something there and was anxious to know what it was without advancing any nearer. William remained still and watched it through the bush. It was big and very excited, and they watched each other for what seemed like ages. He began to feed again and William remained in place for two hours in case he could get nearer to it, watching patiently. As dusk was about to fall William did an imitation of a strange animal crawling (actually walking doubled up) using an umbrella as a tail. The roan was very interested but still kept grazing just out of range. William could have tried a shot once or twice, but he was having too much fun, which he did not want to spoil. Then, to the delight and excitement of his prey, he started to whistle special selections from operas while the animal danced about snorting and at times being singularly vulgar.

At dusk he returned to camp, but when he arrived at the site it was just not there! Could he be wrong about where it was? Thinking about the cook William knew he could do better and had noted some landmarks both near and far. He headed for a bit of high ground and, after an anxious five minutes, was able to verify his position. He reckoned he could not get lost for more than 48 hours at most because, even with no landmarks, the mountains were only 15½ miles (25 kilometres) away giving a north and south line. The dolomite and limestone hills were clear to see and the camps were in the dolomite hills between the two rivers. At worst, he might have got back to a previous camp but that would only mean missing out on dinner and breakfast. Such were William's thoughts as he tried to find his way back; and he arrived just as the inevitable search party was going out to find him.

That evening he did some washing and writing. The carriers arrived back very late and were disappointed not to get any fresh meat issued.

Thursday 28 August

William had selected the next camp site but had too few carriers available to move and carry water because the balance had gone on a three-day hike to look for food. William had a relaxed day of fun. He went out and came across a herd of eland but did not have his gun with him so he sent a man to get it. The eland moved off about a mile away and William stalked after them. Tracking had become difficult because the ground was hard and for the next two hours he had the thrill of tracking as it should be done. He was seldom off-line for more than a few moments and pressed forward at a fast walk. He caught up with the herd twice but the animals were off before he could get a shot at them. They did a lot of dodging and went through some rough bush. He lost them when they were too far from camp but spotted them again a little later some distance further on. By now they did not want to go on because the youngest of the herd was exhausted and could hardly keep up. William found it thrilling and interesting to watch and enjoyed knowing that his tracking skills were improving.

Friday 29 August

The supply of labour began to increase and in the next few days reached 43. He was not supposed to employ more than

35, but it seemed likely that he would have difficulty keeping more than 25.

Saturday 30 August
Today William moved to another perfect camp, except that it earned the name of 'the dusty camp'. There was a lovely view in almost every direction, but the amount of bush made it difficult to get out. There was always a very heavy dew, so it was necessary to be careful when slacking off the tent guys each night. William set about repairing the tent curtains, which involved almost remaking them.

He disgusted his personal staff by placing all the biltong under lock and key so that none could be pinched. The rule was that everyone would be fined if any was stolen and they were more careful after this pronouncement.

Sunday 31 August
On hearing that there was a waterhole only one-and-a-quarter miles (two kilometres) away, William went to investigate. There was water there but even after it had been boiled and filtered it was still unsuitable for drinking, or even for bathing in, about which the local carriers agreed.

William had a queer day. Not feeling up to reading Shakespeare or the newspapers sent from home, he started to read his diaries from 1920 and 1922. He found them interesting to go through again and they made him think a lot about his life. He thought that if the war had come two years later he would still be in the army and that he badly needed knocking into shape but had never properly got to grips with that. Now, though without the comforts and respectability that an army life would have given him, he nonetheless felt happier for it. He thought that he should not have been commissioned as a captain without first being a subaltern and learning the job thoroughly, which would have better suited his temperament. Looking back to 1922 he considered that he had had an easy life, continually dashing up to London for the weekend for a show or concert, which he much enjoyed. These thoughts of the old days only served to make him think critically about himself and he was not happy with his conclusions.

Monday 1 September
William put aside the previous day's contemplations and got off to a hard day's work. He went on a long prospect to the west and just as he was about to return to camp he saw a herd of eland. He finished the job he had allocated himself and then went after the game. It was not easy because the herd was in open long grass, while nearer to him, on the side from which he would have to approach on account of the wind, was a small herd of five roan antelopes. The object at this point was to get between the roan antelope and the eland so that the former would bolt away without disturbing the latter. This objective was achieved but one eland ran round William in a big circle and soon the others were looking to see what the cause of the disturbance was. He had to sit for a very long time while waiting for them to calm down again. A long stalk followed because the herd was still nervous and all the time was walking away slowly, fortunately in an upwind direction. William counted 80 of them (possibly a record sighting), including many very young ones and about 24 large bulls. They seemed never-ending in the distance. Eventually, he got near enough to shoot standing up and hit a bull over the heart. It went off at a steady trot and the remainder lined up and followed in single file. The last round was fired but again he missed.

Tuesday 2 September
William completed the prospecting from the previous day and when he saw vultures went to see if it was his eland, but the eland had got away and only a large dead mamba remained. Later in the day he lit some fires enabling him to locate four roan antelope. He aimed to get beyond them and use his four men to drive them towards him, but twice he failed with his shotgun. On the third attempt the men

started moving too early. The animals could not make William out as he was in white and made no attempt to get near them.

As they went forward in front of him he maintained a steady and fast walk to intercept them without running. It was a beautiful sight. The leader watched William and then started to charge with the other three following and then, rather too late, decided that William was a human and off they all went.

The incident reminded William of a wild horse seeing a jump in a riding school for the first time. First, they went off in a sort of half canter and then suddenly into a wild gallop. At first William felt inclined not to shoot and struggled with his camera, but that was going to take too long. He took a shot at the leader, the second largest animal, with a lethal ball from the shotgun and the animal dropped stone dead, hitting the spine with splinters entering the lungs. He had time to go for the biggest roan antelope, which had beautiful large horns, but having covered it for a moment he decided to leave it.

They were now well off for meat because William had shot a small klipspringer when out on his own as well as a lesser bustard, which is very good to eat. It was difficult to carry the klipspringer because, with too little water left to wash off the dirt afterwards, it could not be carried straight over one's shoulder.

In a day or two the mail arrived, also biscuits and a case of whisky, but ammunition remained short.

Sunday 7 September

William sent off his monthly report to Mills and a note to Blackwell. Then, on Sunday evening, disaster struck. When William first arrived he had had to buy sets of 'chop' boxes that contained rather silly things like Fortnum & Mason's best plum puddings. Although it was only early September William decided to try the first of his puddings. After having been boiled nearly all day in its tin it was served at dinner, but not in the usual manner. Indeed, he found that curried plum pudding with rice, mashed potatoes and papaw squashed up together was not exactly what he had envisaged!

Monday 8 September
Out on his walk William saw the sea again and it appeared to be no more than about three hours' walk away.

Tuesday 9 September
Loads and water had been moved forward again to the next camp but, because of a food shortage and supplies having failed to turn up in the quantities required, they returned to the present site.

One of the carriers turned up looking very sickly after having got lost yesterday while carrying loads forward. He had lagged behind too far and lost sight of the gang. Having spent the night alone in the bush he was terrified, very thirsty and hungry. Luckily for him he was too far from water to be in danger from wild animals.

Wednesday 10 September
William shot an old bushbuck, which took two buckshot and two No. 5 shot before the men could finish it off. He was cut up and hung to dry. The biltong line was now 30 to 40 yards long and the smell never improved.

When an animal is killed nothing is lost except the blood. In the drying process each strip of meat shrinks to between a third and a quarter of its original size. A whole cow would fit into a travelling tin bath without being packed too tightly. It is salted to keep the insects away and the meat is preserved indefinitely.

For the last few days water had been obtained from the River Cubal, which made life a bit easier.

Friday 12 September
William was in camp for most of the day taking care of his sores.

Saturday 13 September
After a long prospect William moved some way forward with more stores to a new site.

Sunday 14 September
This was not to be a good day. Half the men carried one load forward and were then sent off for more water and told to return without fail the next day. The remainder were on short rations and had to do two journeys with loads. The camp was not bad, but it was not easy to get to because for the second half of the route the grass was above the carriers' heads and there was no trace of an animal track. Just beyond the camp there was a path for the water party to use, though they were going forward instead of back, which would have been safer. The local men knew exactly where they were and said that good water was available a short distance ahead. The last water from the Uaque River had been too murky to go through the filter so it had to be boiled.

The carriers did not bless the local guide because it was well after dark before they reached the good water and, having got there, they found only a dry creek bed with no water in the sand. They reached the River Cubal at about midnight.

Monday 15 September
Blackwell turned up for lunch and tea and this was their first meeting for over six weeks. As an honoured guest he was treated to a special lunch, with an extra bird or two, and the chance to use up the 'saved-up' jelly. It was a hot day so they drank the jelly from cups and praised William for his cooking.

Most of the discussion centred on their general discontent with Mills, on how they (Blackwell and William) had had by far the roughest time among the party, and on how they were the only two who could have covered much of the country. Mills was particularly pleased that William had covered the area between the Uaque and Cubal rivers, which he thought would have been almost impossible given the type of country and slow rate of progress. He was also very jealous of William for having seen two herds of eland, for according to Rowland Ward (who had written the latest book on hunting) it was rare to see more than 12 or 15. Having heard this William thought he would send Mills a note about his possibly record-sized bushbuck. The animal shot in July was enormous, but the latest one had horns nearly three inches longer. After taking the meat off the top of the back and a hind quarter for biltong there was still enough to give all the carriers a normal ration to eat.

William's water did not arrive until three in the afternoon and he was very displeased.

After this William left the contact he was supposed to be following and moved nearer water to camp 62. The men were very tired and it was still three hours from the river, but at least the water was clean again. William noted that this area was hard to prospect being broken and hilly and preferred to go back to his previous line and complete a bit he had missed out. However, he had a deadline to meet Mills and there was no time for that.

Tuesday 16 September

It was to be a two-and-a-half-hour walk out and the gang was not ready early. The advantage of this was that it gave the meat time to dry out a little more. The heavy dew had made everything hanging out in the open very wet and needed at least an hour to dry in the hot sun.

The area that remained to be prospected consisted of thick bush and rather steep bluffs leading down to the river. There was unlikely to be anything of interest there so nothing seemed to matter much.

Thursday 25 September

At last William moved down to and across the river, but whether he should have moved on this day was questionable because it was a particularly unlucky day for him. He recalled that in 1915 it had been the battle of Loos, in 1916 he had been wounded in the war, and in 1917 he had been one of the few to get through the battle of Ypres alive.

The River Cubal was said to be infested with crocodiles and now they were having to cross it when it was only a few feet deep. Nothing had been in that area for days, so it was with some fear that they crossed, but without mishap. The camp was near the river and the tsetse flies soon became a menace.

Friday 26 September
William made enquiries to see if water was near the track to the main road, as a rendezvous had been arranged there. He set out to cover about half the distance and observed that he would not like to do the first part of the trip alone, in the dark, unarmed and in the wet season. Even now it was almost a swamp and very dense with an ill-defined track. There was plenty of spoor from bushcow and in wet weather it would undoubtedly be full of crocodiles and snakes.

When halfway across William asked again where the water was, only to be told that there was none until they reached the road. Then he understood why Mattheus returned a day early; it was because he would have had to cross in one day or do without water. Having got across the swamp there was no food for the carriers. Two men had a really hard day going back for food, taking nearly 12 hours in all. They were completely exhausted when they returned, the others were starving and William was furious because it should never have worked out that way.

The camp was set up near the road about three hours away from the rendezvous. William sent a message ahead to say he would meet the lorry near the camp where there were few flies rather than do the extra walk and wait while being bitten by tsetse flies. William remembered a lesson he learnt in the army: it is better to receive a mild reprimand than perform a most unpleasant duty. All that mattered now was that he would soon get his mail at last. The area had been well covered and the few patches that remained were unlikely to yield results, so not worth the time and effort. It was rumoured that the company was going to close down soon and William sent a message to Blackwell to come to lunch to talk about what was going on.

Saturday 27 September
It was an unhappy day for some. A few of the men deliberately disobeyed the *capataz* with the result that six were paid off and two others warned. Two had said they wanted to go some days before, but William waited until it was convenient and so they left today, taking William's mail to Lobito. Mills should have arrived today but failed to do so.

Sunday 28 to Monday 29 September
Monthly reports were written, accounts completed and some more work done on the map, the written work being interrupted by a fruitless three-hour search for roan antelopes. This was also the first time William had seen a road for more than three months and it made him feel positively shy.

Mills finally turned up at dusk this evening, two days late. He camped very close to the local village and sent William a note asking him to go and see him early the following morning. He had brought some of William's mail, which included a box crammed full of goodies including jelly and blancmange.

Tuesday 30 September
William went to see Mills but not very early, only to find that he had not had his breakfast at 7.30 a.m. His excuse was that he had been out in the bush before dawn the previous day, which William found hard to believe.

They avoided conflict and managed to communicate in a civilized manner in the course of which William was told that he had to rush things for a bit and go to a place called

Binga. He was told it was swamp infested with mosquitoes but that there were new varieties of game in the area.

Mills was surprised when William asked for only half the amount of cash he had expected, but even this left him short. As mentioned earlier, though it was customary to pay the men at the end of the month for their previous month's work, William chose to pay his men at least twice a month. Then, if the final payment was paid a bit late it could come out of the next month's funds and by this means William could manage his finances better.

After Binga William thought that he would drop his local labour and head for higher ground, which would be healthier but probably not as good for shooting or work. After that the future was uncertain. He would have completed a year of his contract. The team might stick together and go on to some other concession, or they might be given the option of staying on or going home. William was uncertain whether he wanted to go to the company's new project, which was only just up and running, or go on another prospecting venture. He enjoyed prospecting in situations where he was left alone to get on with his work without interference from others.

After Mills left William finished his reports and went backwards, nearly as far as camp 65, feeling that six days had been wasted when a note would have sufficed. The messengers sent to Lobito had not turned up so he was missing his mail, ammunition and food. William sent out men to find those who were lost and was annoyed that Mills did not bring them in his almost empty lorry. His excuse was that the men had set off three days before he did. They turned up eight days after setting off with more (now surplus) whisky and more labour from Lobito.

Thursday 2 October

William sent a number of men ahead with the *capataz* to prepare a new camp about two hours' walk on. They were to go in an easterly direction because the bush looked bad.

They did not go far enough to the east and soon got into the dolomite hills, returning late in the afternoon completely exhausted and with Mattheus saying that they had not made the new camp because they had got lost. Once in the hills it was impossible to get a sense of direction.

Friday 3 October

William sent for some food and then went out to find the next camp. He struggled to cut his way through the bush and it took him two hours to cover two-and-a-half miles (four kilometres). He found an easier way to return and improved it still further by lighting fires and burning off large stretches of grass, which he could do without fear of hurting the animals because there seemed to be so few of them around.

William had received his ammunition and in the afternoon went out after a herd of bushcows. These are considered the most dangerous game and William had not yet shot one. First he saw a cow and then a youngster, and soon there were seven. Two bulls looked singularly nasty and appeared very near William as if by magic, gazing fondly at him from about 100 yards away, each face looking about as ugly as it could. These animals are hard to kill and if shot they usually charge. It was a nasty moment and William anxiously looked around for a tree to climb. There were hardly any trees about but he was lucky to have one near him and up he went with his rifle. It was not an easy tree to climb and his attempt to take his camera up failed. The first shot hit the big bull right in the chest, near the bottom of his neck from directly in front (supposed to be the best shot). At this point they should have charged but instead of circling the tree they went the other way. The hit animal was barely able to totter after the others. William noted where it went out of sight over the crest of a small hill. He decided it was too dangerous to follow them now, but out of curiosity he went across to see the blood spoor and mark it for tracking the next day. Suddenly the wounded bull and another bolted from just in front of him, but William decided he had urgent things to do in camp and made his way back in open ground with more than one glance over his shoulder. He was unable to trace the tracks of the animal again, although he would like to have had the head as a trophy.

Saturday 4 October
On Saturday and Tuesday William moved forward and on the way got his big bushbuck. He was now on the road for the next four moves forward. It seemed curious to be able to walk without lifting his knees to the level of his chin at almost every step. There was still a water shortage, but knowing that there was some available only 21½ miles (35 kilometres) away by road was easy after the last trials.

Wednesday 8 October
Sending mail had become difficult because there were insufficient men to spare to send to Lobito and it had become almost impossible to find time to write. It was now possible to post letters in Novo Redondo, from where there was a weekly service to Lobito and then the normal service home.

This brings us to the end of William's travel notes taken from his letters home. He then received the following:

10 October 1930
A hand-written note from Mr S. C. Mills (supervising engineer), saying:

> I have had a telegram from London calling the whole party home. Will you therefore move on to the Novo Redondo road, pay off your labour, and wait until I can pick you up there, which should be about the 14th or 15th. Give your labour any spare food to see them to their homes but any surplus do not trouble to bring to Seles.

Mr F. C. Studt for Sir Robert Williams & Company, gave the following reference, dated 12 November 1930:

> To whom it may concern.
>
> This is to certify that Captain W. J. Chads has been in the employ of the Zambesia Exploring Company from January 1930 until November 1930.
>
> During the above period Captain Chads was engaged in general prospecting work in Angola (Portuguese West Africa) under the supervision of our Engineer Mr S. C. Mills, and his work was carried out to the satisfaction of the engineer.
>
> Owing to circumstances over which the Company had no control, the prospecting work had to be stopped and the prospecting party returned home.

Epilogue

William settled down back in England and married Mary D'Olier on 12 December 1931. At first he lived with his wife in a rented house in Nailsworth, Gloucestershire, where he had a son followed by twin daughters. In 1935 he bought his first house, which was in Surrey, and he remained there for the rest of his life. He enjoyed his children, but, despite having a lovely new home, it was a difficult time for him. His mother had just died, he was short of money and it proved to be impossible for him to get back into the army, which is what he had wanted to do, on account of his health. It was not long before circumstances left him bringing up his twin daughters on his own. He was an excellent father who provided a loving and caring home for his daughters, doing everything he could to enable them to stand on their own feet, with the ability to enjoy sports, make music, be creative in the garden and enjoy the countryside.

It was not long, however, before the country was at war with Germany and his prospecting experiences helped him to be remarkably self-sufficient. He kept chickens for eggs and meat, grew fruit and vegetables, tinned his own fruit, made jam, and pickled eggs. He made the most of what was available. He would use table cloths for sheets, turn sheets (by cutting them in half and sewing them together again with the outside edges in the middle) to make them last longer, mix butter with margarine, and desalt butter to make cream.

The country gradually got back on its feet again and William enjoyed a peaceful retirement doing his own cooking and gardening. He would also make regular trips to London to attend concerts and operas, and he played tennis until he was 70. He died in 1969 assured that his work had been completed, Beryl was happily married and Diana was in a good and secure job with British Airways.

Index